CL:AIRE TECHNOLOGY DEMONS 3

DESIGN, INSTALLATION AND PERFORMANCE ASSESSMENT OF A ZERO VALENT IRON PERMEABLE REACTIVE BARRIER IN MONKSTOWN, NORTHERN IRELAND

Paul Beck MSc
Nicola Harries MSc CGeol
Robert Sweeney PhD

Contaminated Land: Applications in Real Environments
(CL:AIRE)

Date: November 2001

CL:AIRE
7th Floor
1 Great Cumberland Place
London
W1H 7AL

Tel: 020 7723 0806
Fax: 020 7723 0815
Web Site: www.claire.co.uk

DESIGN, INSTALLATION AND PERFORMANCE ASSESSMENT OF A ZERO VALENT IRON PERMEABLE REACTIVE BARRIER IN MONKSTOWN, NORTHERN IRELAND

Paul Beck; Nicola Harries; Robert Sweeney

Contaminated Land: Applications in Real Environments (CL:AIRE)

TDP3 © CL:AIRE ISBN 0-9541673-0-9

Published by Contaminated Land: Applications in Real Environments (CL:AIRE), 7th Floor, 1 Great Cumberland Place, London W1H 7AL.

Publication of this report fulfils CL:AIRE's objective of disseminating and reporting on remediation technology demonstrations. This report is a detailed case study of the application of Permeable Reactive Barrier (PRB) technology based on specific site conditions at Nortel Networks' facility in Monkstown Northern Ireland, prepared from a variety of sources. It is not a definitive guide to the application of PRBs. CL:AIRE strongly recommends that individuals/organisations interested in using this technology retain the services of experienced environmental professionals.

EXECUTIVE SUMMARY

The Monkstown site has been operational since 1962 in the manufacture and assembly of electronic equipment and was purchased by Nortel Networks in the early 1990s. Soil and groundwater contamination consisting predominantly of trichloroethene (TCE) and its degradation products dichloroethene (DCE) and vinyl chloride (VC) was discovered during due diligence environmental investigations. Although there was no regulatory requirement to remediate the site at the time, Nortel Networks undertook a voluntary cleanup which consisted of excavation and landfilling of contaminated soil and the installation of a zero valent iron (Fe^0) permeable reactive barrier (PRB) system to treat shallow groundwater in an area of the site known as the eastern car park.

The geology at the site consists of more than 18 m of superficial deposits overlying fine to coarse-grained Sherwood Sandstone bedrock of Triassic age. The drift is characterised by a complex succession of stiff, red-brown clayey till, with intercalated, discontinuous lenses of silts, sands, gravels, and peat, overlain by approximately 0.1 to 1.1 m thickness of made ground.

Shallow water tables occur at depths ranging between 0.45 and 7.82 mbgl. Shallow, horizontal groundwater flow in the vicinity of the eastern car park is interpreted to be in an easterly to northeasterly direction. Calculated hydraulic conductivities range from 3×10^{-6} metres per second (m/s) in coarse silty sand to 5×10^{-9} m/s in clay.

During site characterisation, concentrations of TCE in soil were found to range from 0.3-1,000 µg/kg. Highest concentrations of TCE in groundwater were orders of magnitude greater than other contaminants, with values up to 390,000 µg/L suggesting the presence of free phase TCE.

Laboratory scale feasibility studies, involving column tests on samples of groundwater taken from the site, were used to help design the remedial scheme. The tests showed that TCE reacted very rapidly (half lives of 1.2 to 3.7 hours) with Fe^0, generating cis-1,2-Dichloroethene (c-DCE) as an intermediate degradation product with calculated half lives ranging between 12-24 hours. The column test demonstrated that a significant plume of dissolved iron would be expected to occur downgradient from the PRB resulting in the potential precipitation of siderite ($FeCO_3$) and iron oxide (Fe_2O_3).

A conceptual model of the site hydrogeology, developed by Golder Associates during the site characterisation programme, was simulated using the two dimensional, finite difference, steady state groundwater flow model FLOWPATH. The purpose of the groundwater flow model was to assist in the design of the PRB system and to give additional confidence that the system would operate as designed. The results of the modelling exercise provided an order of magnitude estimation of system parameters and supported the viability of a PRB design at the site. The model demonstrated that the hydraulic regime at the site would not be adversely affected by the installation of a PRB system and that contaminants would not be diverted around the cut-off wall.

Based on field observations, laboratory experiments and modelling, it was decided that the Fe^0 PRB system would be placed in the eastern car park at the property boundary. A cement bentonite cut-off wall would be installed to funnel contaminated groundwater to a vertically aligned underground reaction cell containing Fe^0. It was recognised that some levels of historic TCE contamination remained in subsurface locations downgradient from the proposed PRB installation.

Following the installation of the PRB, a groundwater monitoring programme was established to verify whether the system was operating as designed. The monitoring programme consisted of water level readings and geochemical sampling. Water levels were measured to ensure that the PRB system had not adversely affected groundwater flow conditions. Continuing geochemical monitoring of groundwater upgradient, within and downgradient of the reactive cell, is used to demonstrate that discharge from the reactive cell meets design criteria.

The major ion chemistry shows the predominant groundwater type upgradient of the PRB to be 'calcium bicarbonate'. Groundwater passing through the reactive cell changes from 'calcium bicarbonate' type to 'magnesium-sodium sulphate-chloride', type indicating loss of calcium bicarbonate through calcite precipitation. Contaminant concentrations of TCE are progressively removed as the groundwater flows down through the reactive cell.

Significant decreases in TCE concentrations in some upgradient wells can be explained by: (i) the removal of highly contaminated material during excavation of the PRB and cut-off wall, although some contaminated material remains; and/or (ii) the tail end of a slug of contamination that moved through the site. The degree to which natural variation, natural attenuation, seasonal fluctuations and disturbance during drilling/excavation affect TCE concentrations in wells cannot be determined from the existing data.

Monitoring wells downgradient of the PRB exhibit detectable concentrations of TCE and DCE due to historic contamination. Until such time as levels stabilise and reduce to levels below those found in upgradient wells, downgradient monitoring wells cannot be used to confirm capture of the contaminant plume because their concentrations could mask transgressions of the plume through the cut-off wall.

Monitoring of water levels within the reaction vessel itself indicates periodic reversals of groundwater flow across the reactive cell, making groundwater flow through the reaction vessel difficult to quantify. Estimates, based on potential capture by the cut-off wall and using hydraulic parameters derived from other areas of the site, suggest maximum flow rates of between 1-6 m^3/day, with a residence time within the reaction vessel of between 17.4 and 105 hours.

Estimates of volumetric flow and residence time, along with observed non-detectable concentrations of TCE and DCE in groundwater leaving the reactor at R1, confirm that the reactive cell is operating as designed and meeting the design criterion concentration of 10 µg/L TCE.

Investigations by Queens University Belfast showed that for the Monkstown site, there was some loss in the reactivity of the granular iron at the entrance to the reactive cell. Mineralogical observations showed the presence of calcite and siderite precipitation on the iron in the entrance sample, which was restricted to a very narrow zone. This is likely to cause a reduction in the reactivity and permeability of the Fe^0 over time. At a distance of 10-40 cm into the reactive cell centre, there is evidence of corrosion and increased surface area leading to an increase in the iron reactivity.

No evidence for significant biological fouling was found within the reactive cell since it was first installed.

The remediation costs at Monkstown using a PRB system were £735,500. This included site investigation costs, excavation and disposal of 500 m^3 of heavily contaminated soil, capital costs of the system and monitoring projected forward to 10 years. The estimated equivalent costs for alternatives are £964,500 for landfilling/pump and treat and £865,000 for containment/pump and treat.

Cost effectiveness of the Fe^0 PRB system was considered in terms of contaminant disposition, installation, ongoing operation, and longevity of the system. The PRB system was less expensive to install, and expended less energy than the landfilling/pump and treat and the containment/pump and treat options. In terms of ongoing operation, the system has no requirements for man-made energy and is considered to have a very high operational cost effectiveness. In terms of system replacement, the longevity of the Fe^0 PRB system at Monkstown is expected to be moderate (at least 10-15 years) for minor replacement (iron) and very high (50 years) for any major component replacement.

ACKNOWLEDGEMENTS

This report was prepared by Paul Beck, Nicola Harries and Robert Sweeney of CL:AIRE from information provided from archived company reports and discussions with individuals who were involved in the remedial works. In many cases original data could not be accessed. CL:AIRE would like to acknowledge the support of Dr Graham Norris of Nortel Networks who allowed the use of site information; Mr Simon Plant and Mr Dale Haigh of Golder Associates (UK) Ltd, and Professor Stephan Jefferis of University of Surrey and formerly of Golder Associates (UK) Ltd who provided specific report information and made themselves available to discuss and review various aspects of the project; Mr Robert Essler of Keller Ground Engineering Limited who provided access to project information and Dr Kayleen Walsh of Queens University Belfast who provided information and discussion relating to the detailed performance assessment and reviewed the final document.

CL:AIRE also wishes to acknowledge the following individuals who reviewed and commented on the draft report:

Eur Ing Andrew Dickinson	Corsair Environmental Consultants Ltd
Mr Edwin Jenner	AstraZeneca Plc
Dr Simon Johnson	Certa (UK) Ltd
Professor Jeremy Joseph	JBJ - Environment
Mr Mark Lipman	Rio Tinto Plc
Professor Phil Morgan	Geosyntec Consultants Ltd
Mr Jonathan Smith	Environment Agency - NGWCLC
Dr Steve Wallace	Lattice Property Holdings Ltd

Nortel Networks would like to acknowledge financial support for the installation of the PRB system in the amount of £153,830 from "European Regional Development Fund - Northern Ireland Single Programme (1994-1999) Environmental Services and Protection Sub-Programme."

CONTENTS

List of Figures

List of Tables

List of Plates

List of Appendices

Abbreviations

AQC	Analytical quality control
btoc	below top of casing
bgl	below ground level
Ca^{2+}	Calcium ion
CDM	Construction design and management
Cl^{-}	Chloride ion
CL:AIRE	Contaminated Land: Applications in Real Environments
1,1-DCE	1,1-Dichloroethene
c-DCE	cis-1,2-Dichloroethene
t-DCE	trans-1,2-Dichloroethene
DCM	Dichloromethane
DNAPL	Dense non aqueous phase liquid
EC	European Commission
Fe^{0}	Zero valent iron
FID	Flame ionisation detector
GC	Gas chromatograph
GC/MS	Gas chromatography/mass spectrometry
GGBFS	Ground granulated blast furnace slag
HDPE	High density polyethylene
H&S	Health and safety
HSP	Health and safety plan
ICPS	Inductively coupled plasma spectrometry
K^{+}	Potassium ion
maOD	metres above Ordnance Datum
mbgl	metres below ground level
m/d	metres per day
Mg^{2+}	Magnesium ion
mg/kg	milligrams per kilogram
mg/l	milligrams per litre
mm/yr	millimetre per year
Mn	Manganese
m/s	metres per second
MEK	Methyl ethyl ketone (Butanone)
MIBK	Methyl isobutyl ketone (4-methyl-2-pentanone)
Na^{+}	Sodium ion
NATO/CCMS	North Atlantic Treaty Organization/Committee on the Challenges of Modern Society

NGWCLC	National Ground Water and Contaminated Land Centre
NO_3^-	Nitrate ion
OPC	Ordinary portland cement
PCB	Polychlorinated biphenyl
PCE	Perchloroethylene (Tetrachloroethene)
PID	Photoionisation detector
ppb	Parts per billion
ppm	Parts per million
PRB	Permeable reactive barrier
QA/QC	Quality assurance/quality control
QUB	Queens University Belfast
SAGTA	Soil and Groundwater Technology Association
SEM	Scanning electron microscopy
SO_4^{2-}	Sulphate ion
1,1,1-TCA	1, 1, 1 - Trichloroethane
1,1,2-TCA	1, 1, 2 - Trichloroethane
TCA	Trichloroethane
TCE	Trichloroethene
TCM	Tetrachloromethane
UKAS	United Kingdom Accreditation Scheme
USEPA	United States Environmental Protection Agency
VC	Vinyl chloride
VOC	Volatile organic compound
WESA	Water and Earth Science Associates Limited
µg/kg	microgram per kilogram
µg/L	microgram per litre

1 INTRODUCTION

1.1 BACKGROUND

Nortel Networks' Monkstown Facility is located on Doagh Road in the industrial area of Monkstown, Newtonabbey, in the northeast corner of Belfast, Northern Ireland. The site is currently used as an assembly facility for telecommunication systems.

During the period 1993 to 1996, Nortel Networks (Nortel) commissioned several environmental investigations of the site as part of its corporate due diligence programme. The investigations identified several areas of contamination that were noted for action. The area of interest, which is the subject of this report, is the eastern car park, which was found to contain elevated levels of trichloroethene (TCE) in soil and groundwater. TCE is identified as a category 3 carcinogen in the European Union Dangerous Substance Directive and was the contaminant of most concern at the site. It was determined that remediation of contaminated groundwater should be undertaken to prevent offsite migration to adjacent downgradient land.

The decision to remediate was strictly a voluntary action by Nortel, and was not prompted by any regulatory requirement. Overriding considerations in the choice of remediation technique were: that the remediation would not worsen environmental conditions, that the system would have zero external energy impact, and that no one would be put at avoidable risk during the installation or operation of the system.

An assessment of possible remediation options led to the selection of a zero valent iron (Fe^0) permeable reactive barrier (PRB) as the preferred solution. The PRB was designed in 1994 - 1995 and was installed between November 1995 and February 1996. A groundwater monitoring programme was initiated to assess the performance of the PRB system and is ongoing. The work was carried out by Golder Associates (Golder), who acted as Nortel's environmental consultant. In 1999 Queens University Belfast (QUB) was commissioned to undertake a detailed performance assessment involving the mineralogical study of cores extracted from the reactive barrier itself and carbon isotope analysis of groundwater.

1.2 PURPOSE AND OBJECTIVES

This report is an overview document of Fe^0 PRB technology aimed at those individuals with an interest in the use of cost effective and low energy technologies to manage contaminated land.

The purpose of the Monkstown PRB was to remediate a chlorinated solvent groundwater plume using a cost effective remediation technology with low level maintenance requirements.

The purpose of this report is to describe the environmental conditions that led to the installation of the Fe^0 PRB system and to provide an objective assessment of the performance of the technology. Specific objectives include:

- Describe site characteristics including the nature and distribution of contaminants
- Describe the design and installation of the PRB system
- Provide assessment of current and potential future performance of the PRB system, and

- Provide system costs and cost comparisons with alternative remediation options.

1.3 SCOPE OF WORK OF THE REPORT

The scope of work involved in meeting the objectives for this report included:

- Review of site characterisation studies carried out by Water and Earth Science Associates Ltd (WESA) and Golder Associates
- Review of design elements conducted by Golder Associates
- Review of installation procedures reported by Golder Associates
- Review of monitoring data conducted by Golder Associates
- Review of assessment techniques conducted by QUB, and
- Summary of cost of site characterisation, PRB installation and monitoring.

2 BACKGROUND TO ZERO VALENT IRON PERMEABLE REACTIVE BARRIER DEVELOPMENT

2.1 INTRODUCTION

This chapter provides a summary of the development of zero valent iron (Fe^0) PRB technology. Additional discussion can be found in USEPA (1995) and (1999), Gavaskar *et al.*, (1998), NATO/CCMS (1998) and Powell & Associates (2001).

2.2 WHAT IS A PERMEABLE REACTIVE BARRIER SYSTEM?

A PRB is an *in situ* passive treatment system for remediation of contaminated fluids including groundwater and soil gas. In the case of groundwater treatment, it consists of a permeable wall of reactive material which is installed across the flow path of contaminated groundwater. As the groundwater flows through the permeable barrier, it comes into contact with the reactive material and, depending on the nature of the reactive material, the contaminant is mineralised, degraded to non toxic compounds, or immobilised.

There are various configurations for PRBs. Simple systems involve a continuous reactive barrier. More complex designs employ low permeability slurry bentonite cut-off walls to contain and direct groundwater through a strategically placed reactive barrier. These complex systems are commonly referred to as Funnel-and-Gate™ systems, where the cut-off wall acts as a funnel and the reactive barrier is the gate. See Figure 2.1 for details.

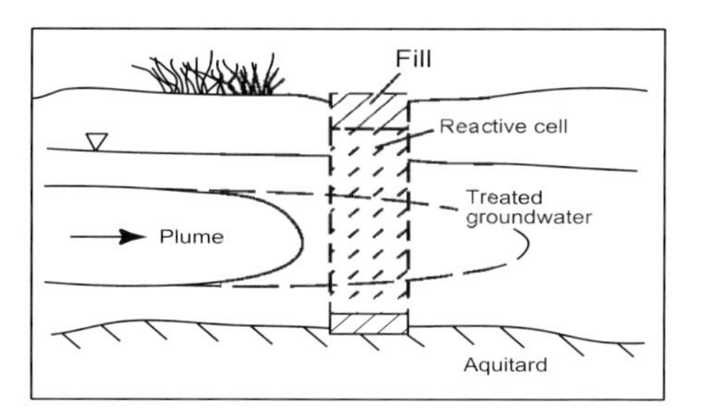

(a) Elevation view of a permeable barrier

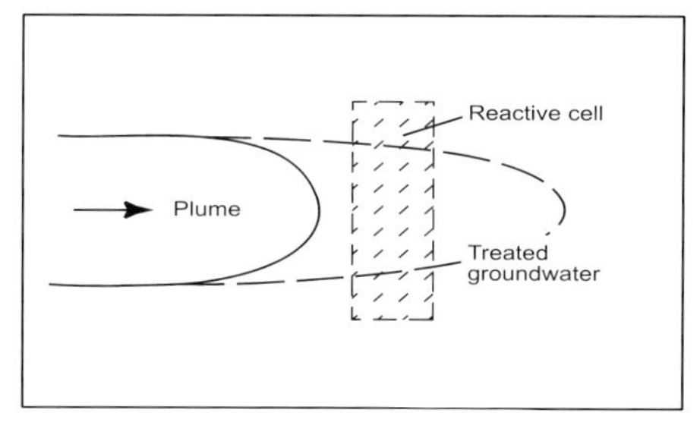

(b) Plan view of a continuous reactive barrier configuration

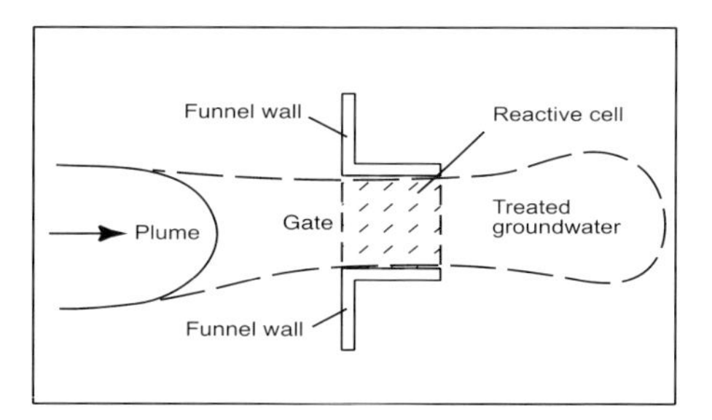

(c) Funnel-and-gate™ system (plan view)

Figure 2.1: Schematics of a permeable barrier configuration (From Permeable Barriers for Groundwater Remediation by A.R. Gavaskar *et al.*, 1998. Battelle Press. With permission.)

The type of reactive material that is used within a PRB will vary according to the type(s) of contaminant(s) to be treated. Fe^0 and other metals/minerals, and carbon substrates such as peat and molasses are used to induce reactions that are redox sensitive. Activated carbon or zeolites are used for sorption. Nutrient delivery systems are used to facilitate bioremediation.

2.3 DEVELOPMENT OF ZERO VALENT IRON PERMEABLE REACTIVE BARRIERS

The use of PRB technology is relatively new. The first Fe^0 PRB was installed in Sunnyvale, California in 1995. The Monkstown installation was the first Fe^0 PRB in Europe. Fe^0 PRBs have now been installed at approximately forty-six sites in USA, one site in Germany, three in Denmark, one in Australia and one in Northern Ireland. Of the fifty-two installations approximately twenty-nine are full scale systems such as at Monkstown and the rest are smaller pilot-scale trials (S. O'Hannesin, Envirometal Technologies - written communication).

Sweeny and Fischer (1972) first reported the use of metals for treating chlorinated organic compounds and they obtained a patent for the degradation of chlorinated pesticides by metallic zinc. Sweeny (1981a, 1981b) reported the breakdown of a variety of contaminants, including: tetrachloroethene (PCE), trichloroethene (TCE), trichloroethane (TCA), trihalomethane, chlorobenzene, polychlorinated biphenyl (PCB) and chlordane by catalytically active granular iron, zinc or aluminium.

The concept of using iron metal for *in situ* remediation of contaminated groundwater resulted primarily from work completed at the University of Waterloo, Ontario, Canada in 1984. It was observed, during research into the sorption of organic contaminants by different well casing materials, that the concentration of the halogenated hydrocarbon bromoform ($CHBr_3$) was reduced when left in contact with steel and aluminium casing materials. The process was described as reductive dehalogenation, but its full significance was not realised until several years later, when the work was published (Reynolds *et al.*, 1990).

Others (Senzaki and Kumagai, (1988a, 1988b) and Senzaki (1988)) reported that powdered iron could be used for the removal of TCE and TCA from wastewater.

In the early 1990s there was increased research activity into reductive dehalogenation involving the use of granular iron, particularly by the University of Waterloo. Gillham and O'Hannesin (1992) provided a discussion of early experimental results and concepts.

The Subsurface Restoration Conference - The Third International Conference on Ground Water Quality Research, held in Dallas, Texas in June 1992, included two key poster presentations by Waterloo researchers, who were among the first to promote the concept of using granular iron in permeable reactive subsurface barriers.

Publications continued to appear in 1993 as researchers presented their early findings. The following year, two major publications by Gillham and O'Hannesin (1994) and Matheson and Tratnyk (1994) presented evidence for the abiotic, reductive dechlorination of a wide range of chlorinated compounds, which followed pseudo first-order reaction kinetics.

The American Chemical Society "Symposium on Contaminant Remediation with Zero-Valent Metals" held in April 1995, presented forty abstracts from thirty different research groups. By this time a significant body of research had developed, culminating in support by the United States Environmental Protection Agency (USEPA) for the development of an innovative *in situ* technology collaborative research project which included the Monkstown site (USEPA, 1995).

In May 1998, NATO/CCMS held a Special Session on Treatment Walls and Permeable Reactive Barriers in Vienna (NATO/CCMS, 1998). In September 1998, USEPA Remedial Technology Development Forum (USEPA, 1998) produced a report which discussed issues arising from the use of PRB technology. The report provided background on the development of PRB technology and summarised several field installations. In September 1999, USEPA published a report detailing the remediation by PRB and long term monitoring of a United States Coast Guard Support Center in North Carolina (USEPA, 1999).

The University of Waterloo currently holds the patent for the use of zero-valent metals for *in situ* groundwater treatment, and has granted commercialisation rights to Envirometal Technologies, Inc., a company partly owned by the University of Waterloo (S. O'Hannesin, Envirometal - personal communication).

2.4 OVERVIEW OF THEORY

The exact mechanism for abiotic reductive dehalogenation of chlorinated compounds by metals is not fully understood, although it is believed that a variety of different pathways are involved, with certain pathways predominating. TCE is degraded by Fe^0 by two pathways: one involving sequential hydrogenolysis which produces DCE and VC intermediates at up to 5-10% of the original TCE mass and a second involving ß-elimination which degrades TCE directly to ethene and ethane (USEPA, 1998). A schematic diagram showing the abiotic degradation of TCE is provided in Figure 2.2.

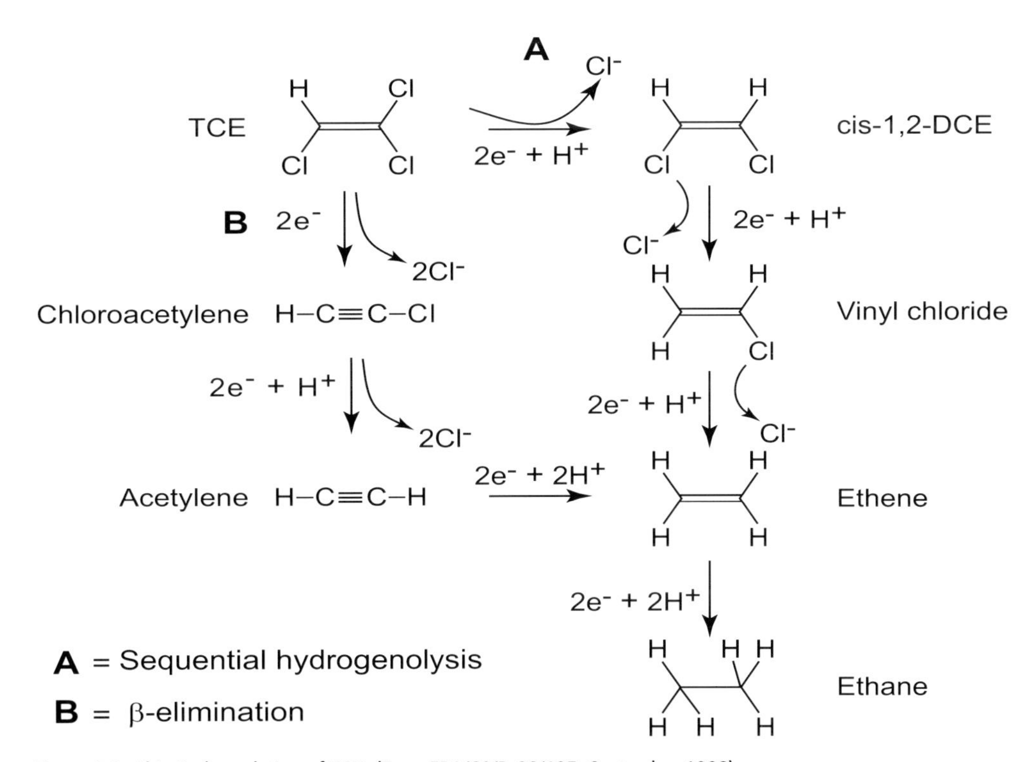

Figure 2.2: Abiotic degradation of TCE. (From EPA/60/R-98/125, September 1998)

The following discussion on reaction chemistry centres on zero valent iron as the reactive metal and a predominantly 'calcium bicarbonate' groundwater to reflect the geochemical conditions at Monkstown.

Under aerobic conditions, free oxygen in groundwater passing through the permeable reactive barrier will react to oxidise the iron and produce hydroxyl ions as shown by Equation 2.1.

$$2Fe^0 + O_2 + 2H_2O \longrightarrow 2Fe^{2+} + 4OH^- \qquad \text{Eqn 2.1}$$

Under anaerobic conditions, where free oxygen is not present in the groundwater, oxygen containing species such as NO_3^-, MnO_2, $Fe(OH)_3$, SO_4^{2-} and CO_2 are reduced by the Fe^0 to form various ionic species.

Under highly anaerobic or reducing conditions, water itself is reduced by iron, to produce hydrogen gas as indicated by Equation 2.2:

$$Fe^0 + 2H_2O \longrightarrow Fe^{2+} + H_2 + 2OH^- \qquad \text{Eqn 2.2}$$

Abiotic reductive dehalogenation of TCE can follow several pathways one of which is shown by Equation 2.3, where TCE is reduced, by the transfer of electrons resulting from the oxidation of iron, to ethene and chloride.

$$3Fe^0 + C_2HCl_3 + 3H^+ \longrightarrow 3Fe^{2+} + C_2H_4 + 3Cl^- \qquad \text{Eqn 2.3}$$

The presence of dichloroethene (DCE) and vinyl chloride (VC) at some sites indicates that intermediate compounds are being produced following the sequential hydrogenolysis pathway.

The hydroxyl ion produced by the reaction described by Equations 2.1 and 2.2 can lead to a rise in pH. As pH increases, bicarbonate ion is reduced to carbonate ion and hydrogen ion as shown in Equation 2.4:

$$HCO_3^- (aq) \longrightarrow CO_3^{2-} (aq) + H^+ (aq) \qquad \text{Eqn 2.4}$$

The rise in pH can be buffered by bicarbonate alkalinity according to the following reaction:

$$HCO_3^- + 2OH^- \longrightarrow CO_3^{2-} + H_2O \qquad \text{Eqn 2.5}$$

The iron and reduced species produced from the reactions between the incoming groundwater and the Fe^0 can react to precipitate a variety of minerals depending upon geochemical conditions. Common reactions include carbonate ion with calcium and iron to form calcite or aragonite ($CaCO_3$); ankerite ($CaFe(CO_3)_2$) or siderite ($FeCO_3$). Iron and hydroxyl can react to form ferric oxyhydroxide (FeOOH) or ferric hydroxide [$Fe(OH)_3$]. A green precipitate, a complex layer of Fe^{2+}, Fe^{3+} with Cl^-, SO_4^{2-} and/or CO_3^{2-} has been observed at some sites.

The above reactions take place rapidly. Under low groundwater flow rates, the length of the reaction zone between the Fe^0 and the groundwater will be relatively short because of the increased contact time between the contaminated groundwater and the Fe^0. The front of the reaction zone itself may migrate up to 5-10 centimetres upgradient from the boundary of the Fe^0 in the reactive cell, because of diffusion, which causes the geochemical effects of the Fe^0 to be displaced (Dr K. Walsh, QUB - personal communication). Under higher groundwater flow rates, the reaction zone will be longer, and the diffusion effects may be decreased or mitigated.

2.5 PERFORMANCE OF PERMEABLE REACTIVE BARRIERS

One of the major concerns surrounding the installation of PRBs is how they will perform over the long term. Performance assessment is related to the design criteria and the remediation objectives set for the site and should consider the effects of groundwater chemistry on mineralogical and chemical reactions, changes in PRB permeability, and the integrity of the cut-off wall.

A comprehensive monitoring programme is necessary for proper evaluation of the performance of a PRB installation. USEPA (1998) emphasised the importance of adequate site characterisation as the first step towards effective monitoring and recommended that performance monitoring cover the following:

- Evaluation of physical, chemical and mineralogical parameters over time
- Verification of emplacement, loss of reactivity, decrease in permeability, decrease in contaminant residence time in the reaction zone and investigation of short circuiting or leakage in the cut-off walls
- Monitoring degradation products, precipitates, hydraulic parameters and geochemical indicators
- Understanding the mechanisms that control contaminant transformation, destruction or immobilisation within the reaction zone

A discussion on assessment criteria used to evaluate the PRB performance at Monkstown is provided in Section 8.

3 SITE DESCRIPTION

3.1 SITE USAGE

The Monkstown facility was purchased by Nortel Networks in the early 1990s. Nortel Networks is a multi-national telecommunication company, which develops and manufactures telecommunication components and internet access systems. The Monkstown site has been operational since 1962 in the manufacture and assembly of electronic equipment. From 1962 until 1985 it was used to manufacture printed circuit boards and assemble electromechanical switching equipment. In 1985 plant operations switched to the manufacture of telephones, switching stations, fax machines and various electronic assemblies.

The site occupies approximately 15 hectares and consists of five large buildings which cover approximately half of the area of the site. The remaining area consists of site roads, car parking and sports fields. The site is surrounded by residential properties to the south and west, and industry to the north and east. Industries in the area include a bus depot and a recently constructed cement works. Figure 3.1 is a regional location map showing the site.

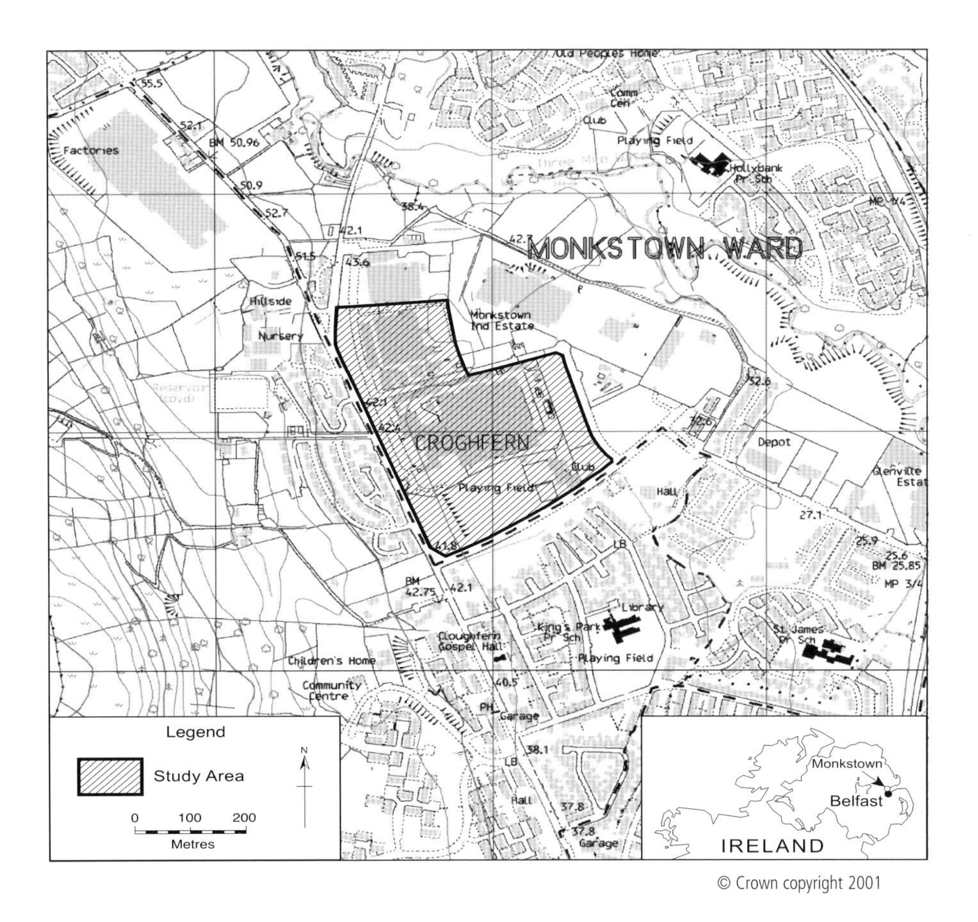

Figure 3.1: Site location plan

3.2 SUMMARY OF ENVIRONMENTAL INVESTIGATIONS AND REPORTS

Field investigations with associated laboratory testing were undertaken to characterise contaminant and hydrogeological conditions at the site and to provide input to the PRB design programme. Early investigations for Nortel were carried out over the entire site as part of its due diligence process for newly acquired property. Later investigations focused on specific problem areas such as the eastern car park where the PRB was installed.

In June 1991, Water and Earth Science Associates Ltd (WESA) undertook a preliminary environmental review of the entire site. This consisted of a walk over tour, the completion of a comprehensive environmental checklist, meetings and interviews with site personnel and a review of available documentation.

In April 1993, WESA conducted a hydrogeological investigation of the site. Twenty-two boreholes were drilled and groundwater monitoring wells were installed. Soil and groundwater samples were recovered and analysed. See Figure 3.2 for borehole locations.

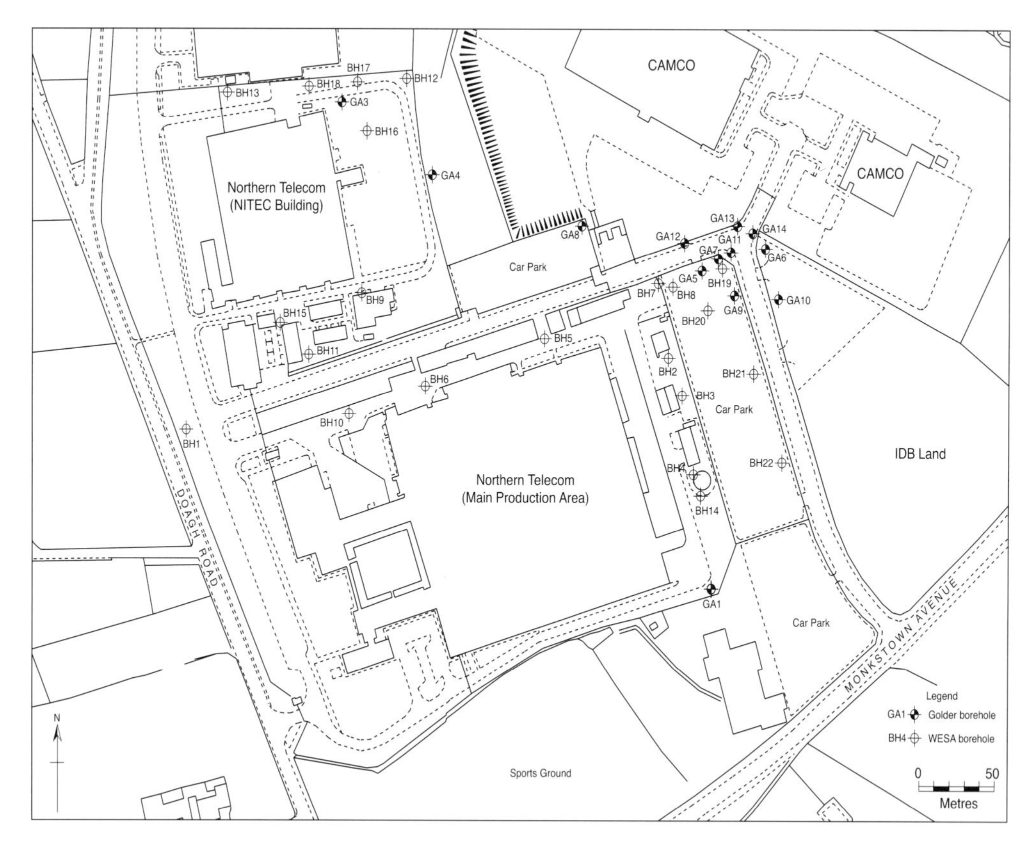

Figure 3.2: Environmental sampling locations Source: Golder Associates (1994)

In 1993 and early 1994, Golder Associates carried out verification sampling of WESA boreholes and groundwater monitor wells and conducted additional intrusive investigations which included the following:

- A shallow soil vapour survey conducted along the northern site property boundary consisting of a total of 33 hand augered holes completed to depths of up to 1.1 metre (m).

- A total of fourteen boreholes to 10 m depth drilled across the site using cable percussion rigs and completed as groundwater monitoring wells with high density polyethylene (HDPE) casing. One round of groundwater samples was collected.

The report included a summary of potential remedial options.

In late 1994, Golder Associates carried out a groundwater flow modelling exercise to assess site hydraulics and to assist in the design of a PRB remedial solution.

In 1995, The Institute of Groundwater Research at the University of Waterloo conducted column treatability tests to assess the suitability of a Fe^0 PRB and to develop design parameters.

In 1995 and up to mid 1996, Golder Associates carried out remedial operations and undertook further site characterisation work and installed five additional shallow groundwater monitor wells in the northern portion of the site away from the eastern car park area.

A summary of environmental sampling locations at the site is shown in Figure 3.2.

3.3 TOPOGRAPHY AND DRAINAGE

Topographic data from 1993 show a maximum elevation difference at the site of 3 m with elevation decreasing from west to east. The site is located on the northwest side of Belfast Lough and is approximately 1.6 km from the water's edge.

The site area receives approximately 1000 - 1200 mm of rainfall annually (Robins, 1994). The majority of the site consists of low permeability building envelopes or paved parking. Runoff is conveyed by drains through an oil/water separator before discharging to a local stream.

3.4 GEOLOGICAL AND HYDROGEOLOGICAL CONDITIONS

3.4.1 GEOLOGY

Within the site area, underlying bedrock consists of fine to coarse-grained Sherwood Sandstone of Triassic age. The depth to bedrock has not been determined at the site but boreholes confirm that superficial deposits consist of more than 18 m of drift overlain by approximately 0.1 to 1.1 m thickness of made ground. The drift is characterised by a complex succession of stiff, red-brown clayey till, with intercalated, discontinuous lenses of silts, sands, gravels, and peat. The thickness of the till and coarser materials varies across the site. The till appears to form a continuous layer from about 10 metres below ground level (mbgl), except along the east boundary of the car park, where the clay till extends from 7.5 to 9.2 mbgl. Below this level, the clay is replaced by a stiff clayey silt, which continues to a depth of at least 10.3 mbgl. In the northeastern corner of the site, adjacent to the PRB, the clay till is commonly less than 0.5 m thick or totally absent. Elsewhere the clay cover is commonly of the order of 3 to 5 m thick.

The coarser materials found in the area are not uniform in texture and vary from clayey silts through silty sands to clean coarse sands and gravels. Laterally these lenses of coarser material appear to thin towards the south and west, whilst in the north and east their extent is less well defined. Cross sections of the superficial geology in the eastern car park area are provided in Figure 3.3. Selected borehole logs are provided in Appendix 1.

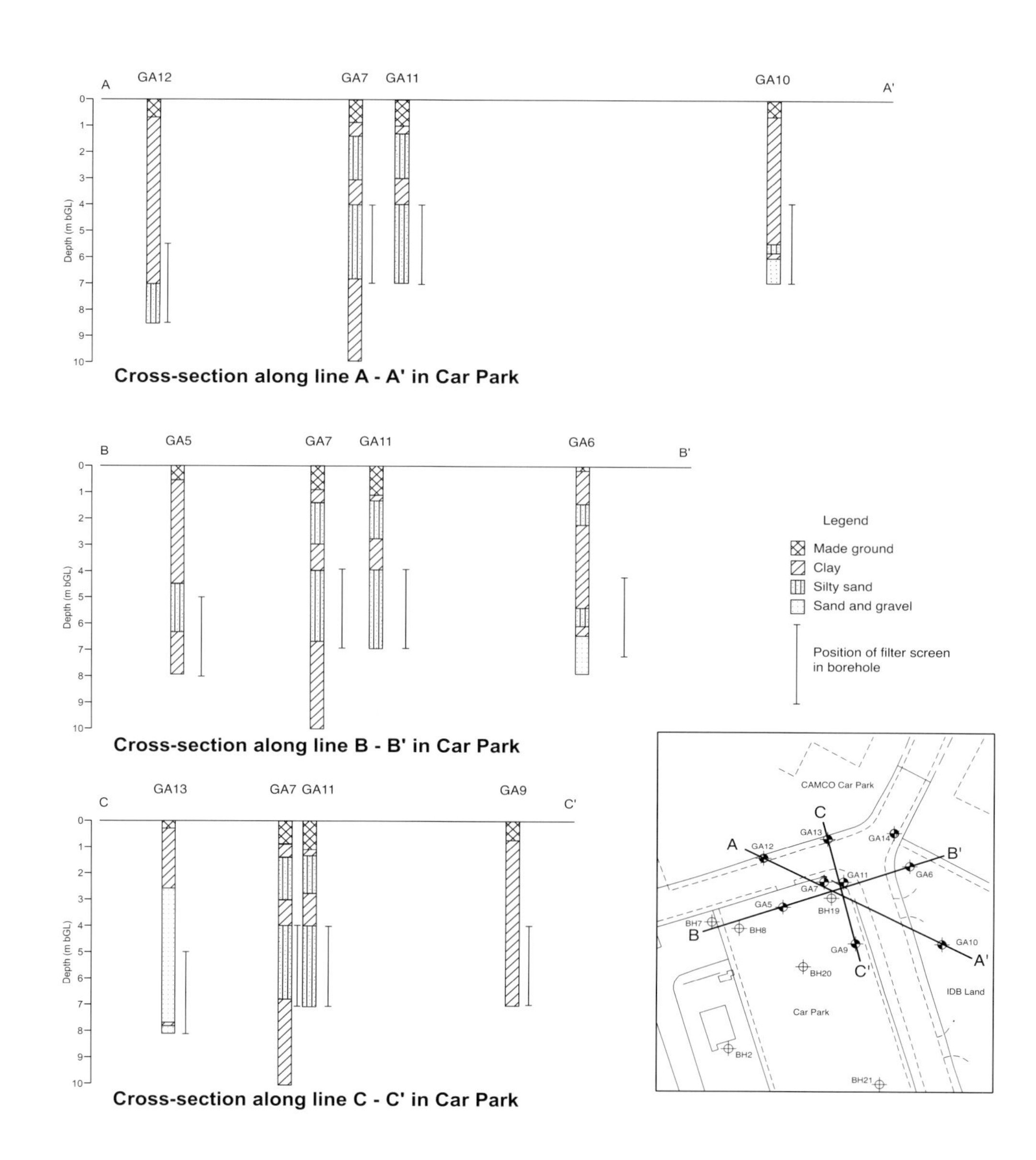

Figure 3.3: Cross section of the geology in the eastern car park area

Source: Golder Associates (1994)

3.4.2 HYDROGEOLOGY

The intercalated and discontinuous lenses of drift give rise to a complex, shallow, unconfined aquifer system within which water tables occur at depths ranging between 0.45 and 7.82 mbgl. A summary of water levels in selected wells taken in March and May 1994, is provided in Table 3.1.

A potentiometric surface map showing the simplified elevation of water tables is provided in Figure 3.4. Horizontal groundwater flow is generally interpreted to be perpendicular to the water table contours. The potentiometric surface reflects the complex geology of the site and anthropogenic influences (building envelopes, drainage, underground cable conduits, roadways etc) but overall, shallow groundwater flow in the vicinity of the eastern car park is shown to be in an easterly to northeasterly direction. The vertical component of groundwater flow was not interpreted.

Table 3.1: Water level data March 1994 and May 1994

BH No	BH Elevation (maOD*)	Date Dipped	Water Level (mbtoc**)	Water Level (maOD*)	Date Dipped	Water Level (mbtoc**)	Water Level (maOD*)
BH2	36.18	18/3/94	2.95	33.23	6/5/94	2.94	33.24
BH7	36.20	-	-	-	6/5/94	5.18	31.02
BH8	36.10	18/3/94	5.59	30.51	6/5/94	5.72	30.38
BH12	40.25	18/3/94	3.31	36.94	6/5/94	3.49	36.76
BH16	40.21	-	-	-	6/5/94	0.89	39.32
BH17	40.22	18/3/94	0.83	39.39	6/5/94	0.89	39.33
BH18	40.22	18/3/94	0.60	39.62	6/5/94	0.72	39.50
BH19	36.10	18/3/94	5.38	30.72	6/5/94	5.40	30.70
BH20	36.15	18/3/94	0.59	35.56	6/5/94	0.62	35.53
BH21	36.10	17/3/94	1.21	34.89	6/5/94	1.30	34.80
GA3	40.20	18/3/94	1.68	38.52	6/5/94	0.70	39.50
GA4	40.40	18/3/94	2.41	37.99	6/5/94	2.64	37.76
GA5	36.10	18/3/94	5.42	30.68	6/5/94	5.64	30.46
GA6	36.50	16/3/94	1.80	34.70	6/5/94	6.70	29.80
GA7	36.10	18/3/94	5.36	30.74	6/5/94	5.43	30.67
GA9	35.80	18/3/94	6.76	29.04	6/5/94	2.11	33.69
GA10	36.10	17/3/94	6.25	29.85	6/5/94	6.29	29.81
GA11	36.30	18/3/94	5.62	30.68	6/5/94	5.64	30.66
GA12	36.30	25/3/94	5.88	30.42	6/5/94	6.10	30.20
GA13	36.30	25/3/94	5.99	30.31	6/5/94	6.13	30.17
GA14	36.10	25/3/94	7.69	28.41	6/5/94	5.29	30.81

* metres above Ordnance Datum
** metres below top of casing

Source: Golder Associates (1994)

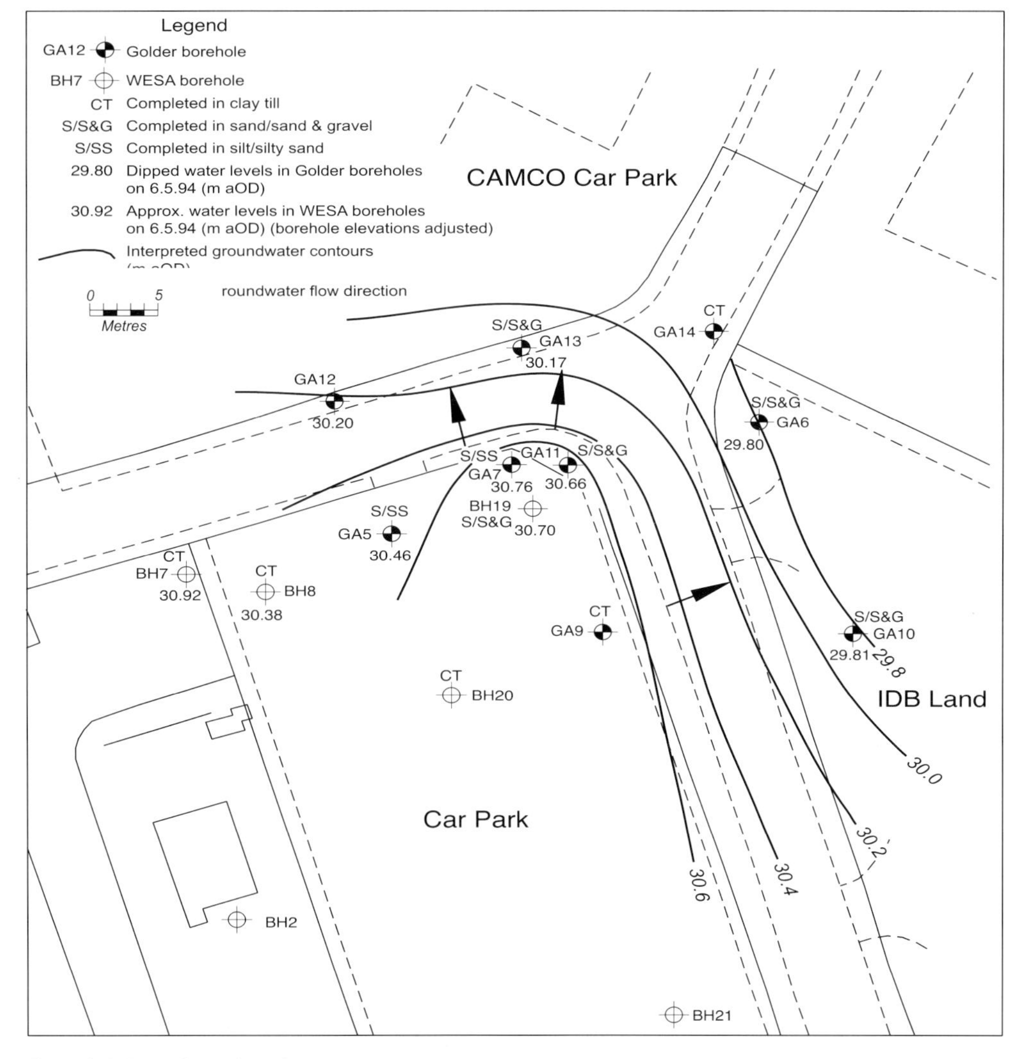

Figure 3.4: Potentiometric surface map

Source: Golder Associates (1994)

Hydraulic conductivity testing was carried out on samples from three wells completed in different textured materials including a sandy clay (GA7), clay (GA9), and coarse, silty, sand (GA11). Testing involved falling head slug tests interpreted using the method of Hvorslev (1951). Calculated hydraulic conductivities are provided in Table 3.2 and ranged from 3 x 10^{-6} metres per second (m/s) in coarse, silty sand to 5 x 10^{-9} m/s in clay.

Table 3.2: Calculated hydraulic conductivities

BH No.	Geology	Hydraulic Conductivity (m/s)
GA7	Sandy clay	1 x 10^{-6}
GA9	Clay	5 x 10^{-9}
GA11	Coarse silty sand	3 x 10^{-6}

Source: Golder Associates (1994)

Major ion chemistry can be plotted on trilinear diagrams such as Piper diagrams, to differentiate groundwater types. A discussion of the use and construction of Piper diagrams is provided in Freeze and Cherry (1979). Monitoring wells located upgradient from the area to be remediated are dominated by 'calcium bicarbonate-sulphate' waters with lesser but significant amounts of magnesium, sodium and chloride in some wells (see Section 7) as indicated by the Piper diagram in Figure 3.5.

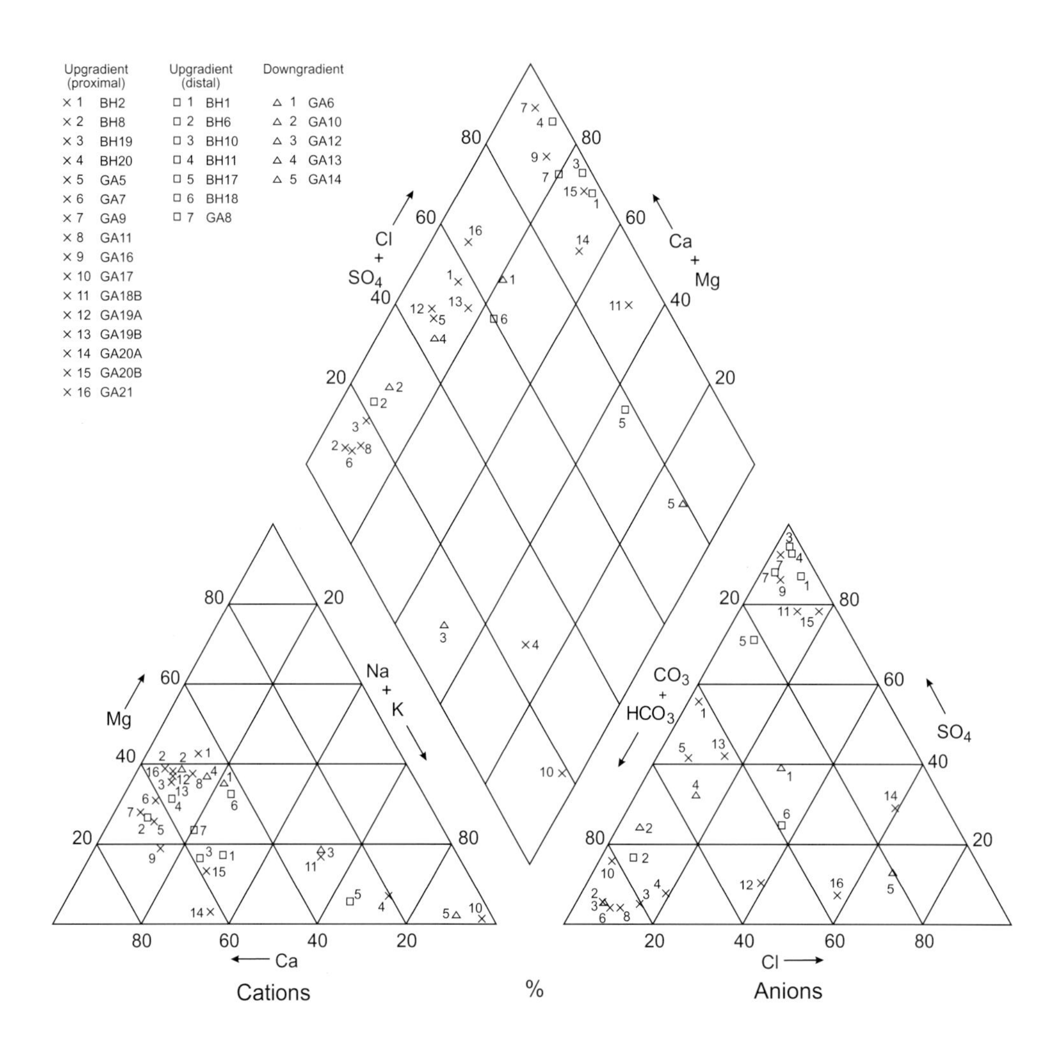

Figure 3.5: Piper diagram showing major ion chemistry of groundwater

Source: Golder Associates (1994)

3.5 NATURE AND EXTENT OF CONTAMINATION

The main driver for remedial action at the site was the presence of TCE.

Soil and groundwater samples from the site were analysed for metals and volatile organic compounds (VOCs). UK statutory remediation criteria for soil and groundwater did not exist for the organic contaminants identified at the site, at the time of the initial environmental investigations. Consequently analytical results from site samples were compared against the Dutch remediation criteria which included both Target Values and Intervention Values (Ministry for Housing, Planning and the Environment, 1994) for both soil and groundwater. These values were used only as an initial guide to assess the degree of contamination and potential environmental risk at the site.

Since it became apparent from the investigation that contaminated groundwater was an issue at the site, the European Drinking Water Standards (Council of the European Community, 1980) were also reviewed with a view to developing site specific remediation criteria.

A summary of soil and groundwater remediation criteria are listed in Table 3.3.

Table 3.3: Soil and groundwater remediation criteria

Contaminant	Soil		Groundwater		
	Dutch Target Values *	Dutch Intervention Values *	Dutch Target Values	Dutch Intervention Values	DWS
PCE	2	200	0.01	40	10 (with TCE)
TCE	0.2	12,000	0.01	500	10 (with PCE)
c-DCE	n/a	n/a	n/a	n/a	30 (WHO)
VC	n/a	0.02	0.01	0.7	0.5
DCM	0.2	4000	0.01	1000	n/a
MEK	n/a	n/a	n/a	n/a	n/a
Acetone	n/a	n/a	n/a	n/a	n/a

*Target and Intervention Values calculated by Golder assuming average soil to have2% organics and 55% clay
All values in soil in µg/L. All values in groundwater in µg/kg.
n/a = No guideline value available. DWS - EU Drinking Water Standard; WHO - World Health Organisation

3.5.1 SOIL CONTAMINATION

Contamination in the soil consisted of a variety of VOCs and selected analyses of these contaminants are provided in Table 3.4. Additional results are available in Appendix 2.

Table 3.4: Selected analyses of soils for VOC contaminants

Borehole no. and sample depth	Total VOCs*	TCE	DCM	MEK	Acetone
GA6 (8m)	74	47	2	20	61
GA7 (5m)	17561	930	1300	15000	4900
GA9 (2m)	24.3	0.3	1	17	59
GA11 (6m)	1098	1000	7	72	89
GA12 (1m)	18.4	1	4	10	45
GA12 (7m)	197	180	4	10	31
GA13 (1m)	51.5	34	4	10	370
GA13 (6m)	1099	990	10	39	85
GA13 (8m)	223.8	210	2	8	40
GA14 (5m)	16.3	0.4	4	10	33

All results in µg/kg (ppb)
*No guideline value available

Source: Golder Associates (1994)

The highest concentrations were found for TCE, dichloromethane (DCM), methyl ethyl ketone (MEK), acetone, and PCE with respective µg/kg (microgram per kilogram or ppb) values ranging from 0.3-1,000; 1-1,300; 8-15,000; 31-4,900 and 0.2-11. Total VOC readings ranged from 16.3-17,561 µg/kg. Levels of TCE, DCM and PCE exceeded Dutch Target Values but were below Dutch Intervention Values. There were no Dutch criteria for MEK or acetone. Metal concentrations were below target values.

The highest readings of total VOCs were found in boreholes GA7 at a depth of 5 m; GA11 at a depth of 6 m and GA13 at depths of 6 and 8 m. The sample from GA7 was taken from a sandy clay while samples from GA11 and GA13 were associated with coarser horizons of coarse, silt, sand and gravels.

3.5.2 GROUNDWATER CONTAMINATION

VOCs were elevated in groundwater. Dutch Intervention Values were exceeded for TCE and PCE only, with elevated levels of other solvents including: 1, 1, 2- trichloroethane (1,1,2-TCA), DCM, chloromethane, methyl isobutyl ketone (MIBK), MEK, acetone and toluene. Highest concentrations of TCE were orders of magnitude higher than other contaminants, with values of 30,000 µg/L in GA7; 43,000 µg/L in GA13; 69,000 µg/L in GA11; 250,000 µg/L in BH19; and 390,000 µg/L in GA19. The concentrations of TCE in BH19 and GA19 are approximately 20 % to 30 % of the aqueous solubility of TCE and suggest the presence of free phase TCE. The source of the high TCE levels is thought to be the previous storage of drums in the car park area. Low concentrations of DCE of 38 µg/L in GA11 and 2 µg/L in BH21, and VC of 8 µg/L in BH21 and 0.1 µg/L in GA9 suggest that limited natural biodegradation of chlorinated solvents may be occurring at the site.

Concentrations of other solvents were generally less than 100 µg/L with the exception of acetone and MEK in GA7, GA11, GA13 and BH19. GA7 is screened in a sandy clay, whereas GA11, GA13 and BH19 are screened predominantly in coarse sands and gravels.

Selected analyses of groundwater samples for VOC contaminants are provided in Table 3.5. More complete results are available in Appendix 2.

Table 3.5: Selected analyses of groundwater for VOC contaminants

	Contaminant						
BH No	**PCE**	**TCE**	**c-DCE**	**VC**	**DCM**	**MEK**	**Acetone**
GA5		**2,400**			22	220	120
GA5 (D)		**1,500**			8	98	130
GA6	**41**	**15,000**			44	410	240
GA7		**30,000**			220	2,100	1,200
GA9		2		0.1			
GA11		**69,000**	38		210		450
GA11 (D)		**50,000**			570	3,900	4,200
GA12	9	**4,900**			44	430	190
GA13		**43,000**			430	4,100	1,600
GA19		**389,932**					
BH19	**84**	**250,000**			430	4,400	1,800
BH21			2	**8**			

All results in µg/L (ppb)
Results in bold show concentrations in excess of Dutch Intervention value
Results are only shown for determinands where detection limits are exceeded
(D) Duplicate sample tested

Source: Golder Associates (1994)

4 TECHNOLOGY DEMONSTRATION SUPPORT ISSUES

4.1 INTRODUCTION

This section discusses supporting issues associated with the selection, installation and post-installation monitoring of the PRB system including:

- Regulatory approval and compliance
- Contract agreement and health & safety
- Work plan
- Sampling plan
- Laboratory analytical methods, and
- Quality assurance/quality control

4.2 REGULATORY APPROVAL AND COMPLIANCE

The site characterisation and remediation was undertaken by Nortel following their own policy of proactive, voluntary clean up. Although contamination of groundwater had been identified at the site, at the time of the works there was no regulatory requirement to undertake remedial action. The Department of the Environment, responsible for environmental regulation in Northern Ireland, were made aware of the findings of the site characterisation and of Nortel's intentions to undertake remediation.

4.3 CONTRACT AGREEMENT AND HEALTH AND SAFETY

Site works were carried out by Golder Associates under contract to Nortel. The Pre-Tender Health and Safety Plan (HSP) for the work was developed by Golder and complied with Nortel's corporate Health and Safety (H&S) requirements.

The installation of the PRB was carried out under Construction (Design and Management) Regulations (CDM)(1994). Principal parties covered under CDM were:

- Client - Nortel Networks
- Planning Supervisor - Golder Associates
- Principal Contractor - Keller Ground Engineering Ltd
- Designer - Golder Associates

A Construction Phase HSP for the installation of the PRB was developed by Keller Ground Engineering Ltd. The plan included risk assessments, details of individual work practices, requirements for personal protection equipment, monitoring, responsible parties and details on general site procedures.

4.4 WORK PLAN

The work plan for the installation of the PRB was developed jointly between Golder and Keller Ground Engineering Ltd and comprised a detailed method statement. Due to the relatively novel remedial approach, a detailed work plan was produced in order to limit the potential for problems occurring during the installation. Particular attention was given to the exact sequencing of the PRB installation as there was a limit to the time during which the reaction vessel could be inserted within the slurry, before the slurry 'set'.

4.5 SAMPLING PLAN

Site characterisation was carried out by WESA and Golder Associates.

The WESA site investigation of 1993 involved the drilling of 22 boreholes, all of which were completed with the installation of piezometers for groundwater monitoring. An HNu HW 101 (11.7 eV) photoionization detector (PID) and a Neotronics EXOTOX 40 multi-gas monitor were used to continuously monitor workers' air space at the top of the exploratory holes during drilling. Soil samples collected during drilling were screened in the field using a PID to measure the general levels of contamination prior to being transported to the laboratory for analysis. Groundwater samples were collected from dedicated check valves and tubing installed within monitoring wells.

The Golder site investigation of 1994 involved a soil gas survey, the drilling of boreholes to depths of between 7 m and 10 m, installation of piezometers in the boreholes and collection of groundwater samples. Re-sampling of groundwater from the WESA boreholes was also undertaken.

Thirty - three shallow holes for sampling soil gas were installed to depths of up to 1.5 m. Portable gas chromatograph (GC) model OVA 128 GC with flame ionisation detector (FID) manufactured by Foxboro Instruments was used to measure soil gas concentrations of volatile organic compounds in the field.

Soil samples were collected every one metre from the 14 boreholes installed by Golder. Headspace analysis was carried out on site using the OVA 128 GC. Samples containing elevated concentrations of volatile organics were selected for laboratory analysis.

Groundwater samples were collected using a dedicated check valve sampling system, consisting of a 25 mm Teflon™ foot valve connected to a 21 mm internal diameter polyethylene tube, following purging of three borehole volumes of water from each borehole. In those boreholes where recovery was very slow, the borehole was purged dry and a sample of the fresh recharge collected.

4.6 LABORATORY ANALYTICAL METHODS

Soil and groundwater samples were analysed by Chemex International PLC (Chemex) and ALcontrol Laboratories (formerly ALcontrol Geochem Laboratories). Both laboratories are accredited under the United Kingdom Accreditation Scheme (UKAS).

Chemex carried out the VOC analysis using a UKAS accredited, modified USEPA purge and trap method of gas chromatography/mass spectrometry.

ALcontrol undertook inorganic analyses using the following UKAS approved techniques for each of the analytes listed below.

Calcium	- Inductively Coupled Plasma Spectrometry (ICPS)
Magnesium	- ICPS
Sodium	- ICPS
Potassium	- ICPS
Iron	- ICPS

Manganese	- ICPS
Alkalinity	- Titration
Chloride	- Ion Chromatography
Sulphate	- Ion Chromatography
Nitrate	- Colorimetric
pH	- Ion selective electrode

4.7 QUALITY ASSURANCE / QUALITY CONTROL (QA/QC)

QA/QC for site characterisation, installation of PRB and post installation monitoring is discussed below.

4.7.1 FIELD QA/QC

Field methods carried out by Golder complied with UK industry best practice at the time as summarised below.

During the site investigation, all equipment that was used for drilling and sampling was steam cleaned prior to mobilisation onto each borehole location to prevent cross contamination. During drilling no water was added to the drilling process and only a restricted amount of vegetable oil was permitted to be applied to equipment as lubrication if absolutely necessary. The steam cleaning was undertaken in a designated area of hardstanding away from the drilling area. Only potable water was used in the steam cleaner. A non-phosphate detergent was added to the water during cleaning of equipment which was subsequently rinsed in potable water. All wash water was collected in a lined sump and analysed for contaminants prior to discharge.

Soil and groundwater samples were collected in laboratory supplied, clean containers following written protocols. Samples were collected by field personnel wearing disposable plastic gloves, and soil sample equipment was decontaminated between samples by rinsing with potable water to prevent cross contamination. Samples were transferred to insulated containers and refrigerated with ice packs for delivery to the laboratory. Chain of custody documentation was completed on site as each sample was transferred to the storage container. All samples were transported to the laboratory within 24 hours of collection.

Replicate samples were taken in the field at a frequency of at least one sample per 50 soil samples and one sample per 20 groundwater samples. All replicated analytical results for groundwater were verified within 10% of each other.

Trip blanks comprising bottles containing de-ionised water were supplied by the laboratory for each sample delivery.

4.7.2 LABORATORY QA/QC

Laboratory analysis complied with industry best practice for analytical methods.

4.7.2.1 Organic Analysis

Samples were stored by the laboratory in refrigerated facilities at 4 ^{o}C and analysed within the specified holding times (7 days maximum).

Samples were routinely spiked immediately prior to analysis with a standard mixture containing three internal standards and three surrogate compounds, in order to permit reliable quantitation and monitoring of recovery efficiency. Acceptable spike recoveries range from 75-120 %. Sample results were provided "as reported" and in some cases results were below the method detection limit. Aqueous samples were subject to dilution if highly contaminated but were otherwise analysed directly without further preparation or treatment.

Soil samples were analysed as a mixture with reagent water, or by methanolic extraction followed by addition to reagent water (on occasions when the target compound concentrations were appropriately high).

Duplicate soil and groundwater analysis were carried out during each sampling round. The chemical results of the duplicates were checked against each other and were always within 10 %.

Target compounds were identified and quantified by reference to a calibration standard previously run within the same analytical period on that instrument. Both soil and groundwater analyses were related to a reagent water blank analysis which was carried out at the start of the period to confirm the absence of contamination introduced within the laboratory.

Data processing was carried out using automated routines and the results obtained were subjected to a series of checks in order to confirm the validity of the automated assignments. The peak areas for the internal standard compounds and the recoveries of the surrogate compounds were also monitored for the internal laboratory quality control purposes. Any significant deviation from general procedures was recorded.

4.7.2.2 Inorganic Analysis

All samples were stored at 5 ± 3 °C unless analysis commences on the day that samples are brought to the laboratory.

Samples are analysed under a strict programme of Analytical Quality Control (AQC). The primary purpose of AQC is the identification of significant changes in the performance of the Analytical System. It is not intended to assign either the cause of the change or magnitude of the effect upon the affected samples. This method of monitoring performance has a reasonably high probability of detecting significant changes with respect to within batch and between batch random errors and systematic errors (bias).

The preparation of control charts is carried out to derive mean (target) values from a known target concentration. Standard deviation values are set up using a fixed 2.5 % RSD.

Several control charts (showing an AQC at different concentrations) may be set up for a single determinand, the requirement for number of Quality Control samples is defined in the methodology for each analytical method.

Control charts are reviewed on a routine basis (a formal check shall be performed every 3 months) with regard to breaches of ± 2 and ± 3 times the standard deviation from the target value and with respect to any trends that are seen in the derived data. This duty is performed by the Quality Section in association with the Senior Analysts or other appropriate representatives from the Laboratory Sections. The action reported in response to breaches of two and three standard deviations from the mean value shall be monitored by the Quality Section and / or the Principal Chemists. Note, AQC failures should be expected at approximately 1 in 20 for breaches of $\pm$ 2 standard deviations and 1 in 100 for breaches of $\pm$ 3 standard deviations.

A control chart shall be maintained for plotting individual AQC values, this shall detail all individual values and shall plot all data points for comparison against warning and action limits. The number of control charts used for recording individual AQC values are defined in relevant Tables.

4.7.3 PERMEABILITY OF SLURRY WALL QA/QC

To maintain quality assurance on the permeability of the wall, routine checks were conducted on the material properties of the liquid slurry during construction. Samples of the liquid slurry were also taken and allowed to cure before being transferred to Laing Technology Group (LTG) Services Laboratories for testing for permeability, strength and liquidity.

5 REMEDIATION DESIGN

5.1 INTRODUCTION

Three distinct phases were built into the remediation design. They include:

- Laboratory scale feasibility study
- Numerical model and conceptual design of the cut-off wall and reaction vessel
- Design of the cut-off wall and reaction vessel

5.2 LABORATORY STUDIES

Laboratory scale feasibility studies involved column tests, which were carried out by the Institute for Groundwater Research, University of Waterloo, using samples of groundwater taken from the site. The following sections were taken from a report prepared by the Institute for Groundwater Research (1995).

5.2.1 OBJECTIVES

The objectives of the column tests were:

1. to determine whether the chlorinated organic compounds would degrade as they passed through the reactive granular iron under potential flow conditions at the site, and
2. to establish parameters for the detailed design of the PRB.

5.2.2 METHODOLOGY

A reactive column experiment was set up using 100 cm long and 3.8 cm diameter Plexiglass™ columns packed with granular iron. Seven sampling ports were positioned along the length of each column at distances of 5, 10, 20, 30, 40, 60 and 80 cm from the inlet valve as illustrated in Figure 5.1.

The packed column was flushed with carbon dioxide to avoid air entrapment, then flushed with distilled water. Contaminated groundwater from the site was introduced from a Teflon bag and pumped to the bottom influent end of the column. The column was periodically sampled over time until a steady state concentration profile was achieved. A 1.5 or 2.0 mL sample was collected from each sampling port and tested for halogenated and non-halogenated volatile organic compounds, redox potential and pH. Samples were also obtained from the influent solution and effluent overflow bottles and tested for the same analytes as well as inorganic constituents.

Flow velocities of 109 cm/day and 54 cm/day were used in the study.

5.2.2.1 Analytical Procedures for Organic and Inorganic Analytes

Analyses for the less volatile halogenated organic compounds were carried out using a pentane extraction procedure and a gas chromatograph equipped for electron capture detector. The more volatile halogenated organic compounds were analysed using headspace analysis and a gas chromatograph

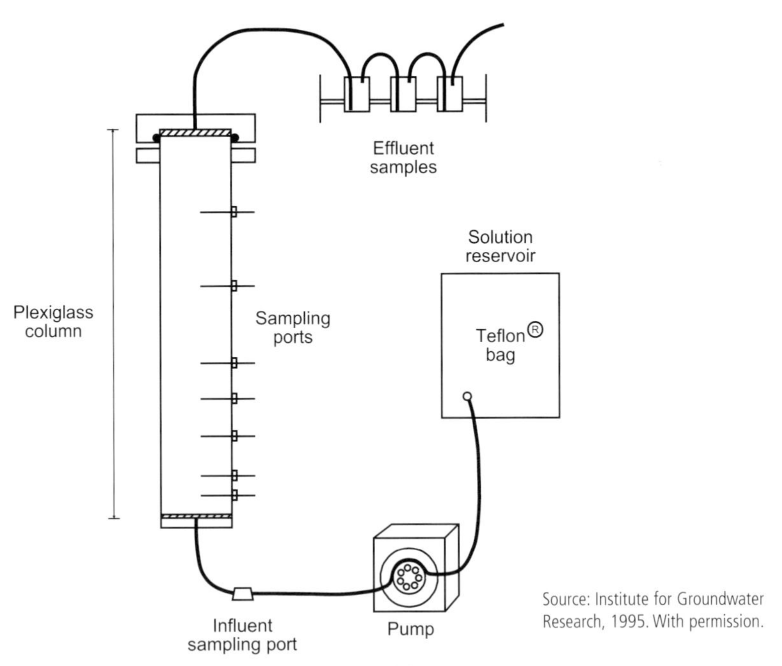

Figure 5.1: Schematic of the apparatus used in the column treatability experiments

equipped with photoionisation detector. Non-halogenated organic compounds were analysed using a gas chromatograph equipped with a flame ionisation detector. Detection limits for all compounds were determined using the USEPA procedure for Method Detection Limit.

Redox potential was determined using a combination Ag/AgCl reference electrode and meter. A pH/reference electrode and meter were used to measure pH.

Cation analyses for sodium (Na^+), magnesium (Mg^{2+}), calcium (Ca^{2+}), potassium (K^+), total dissolved phase iron (Fe) and manganese (Mn) were undertaken by atomic absorption spectrophotometry.

Anion analyses for chloride (Cl^-), sulphate (SO_4^{2-}) and nitrate (NO_3^-) were by ion chromatography.

Alkalinity was measured by titration.

Dissolved oxygen was measured using a modified Winkler titration method.

5.2.3 RESULTS

5.2.3.1 Organic Contaminants

The major organic contaminants detected in the groundwater at the Nortel site were TCE, PCE, 1,1,2-TCA, c-DCE and MEK. Concentrations of c-DCE were difficult to interpret due to coelution with MEK, but because MEK is known not to degrade or be produced in the presence of iron, any overall changes in concentration were attributed to the production of c-DCE. Trace amounts of 1,1,1-trichloroethane (1,1,1-TCA), tetrachloromethane (TCM), trans- 1, 2-dichloroethene (t-DCE), 1, 1-dichloroethene (1,1-DCE) and VC were also detected.

TCE, PCE and 1,1,2-TCA declined to non-detectable concentrations within the column. Concentrations of c-DCE and MEK initially increased due to the dechlorination of TCE and then decreased along the remaining portion of the column. Steady-state decay curves for TCE and c-DCE and MEK for the faster flow velocity are shown in Figure 5.2. Trace amounts of 1,1-DCE, t-DCE and VC were observed in the column and were attributed primarily to the dechlorination of TCE, with lesser contributions from PCE and DCE isomers. Neither 1,1,1-TCA nor TCM were detected in the column.

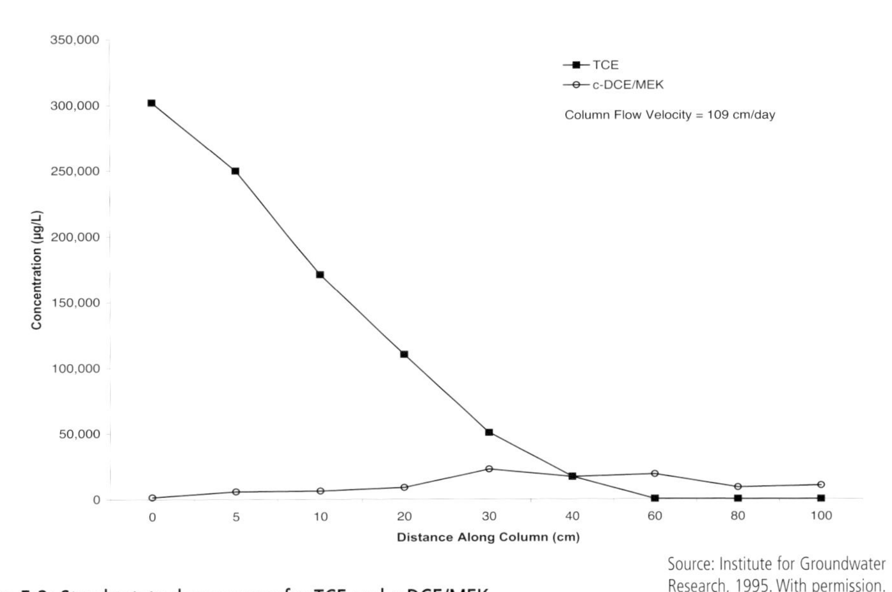

Source: Institute for Groundwater Research, 1995. With permission.

Figure 5.2: Steady-state decay curves for TCE and c-DCE/MEK

Steady-state concentration profiles were plotted for the two different flow velocities for TCE, PCE, 1,1,2-TCA and c-DCE/MEK. The exponential decline of contaminant concentration versus distance along the column indicates that degradation followed first-order reaction kinetics. Converting distance along the column to time using flow velocity, degradation rate constants can be calculated using the first-order kinetic model:

$$C=C_0e^{-kt} \qquad \text{Eqn 5.1}$$

Where,
C is the organic concentration in solution at time t
C_0 is the initial organic concentration (i.e. organic concentration in the influent solution)
k is the first-order rate constant
t is time

Equation 5.1 can be re-arranged to:

$$\ln (C/C_0) = -kt \qquad \text{Eqn 5.2}$$

The compound half life, where $C/C_0=0.5$, is the time at which the initial concentration declines by half and can be calculated by rearranging Equation 5.2 to:

$$t_{1/2} =0.693/k \qquad \text{Eqn 5.3}$$

The slope of the first-order kinetic model was used to calculate the decay constants. The coefficient of variation (r^2), which is a measure of how well the experimental data fit the first-order decay equation, shows a strong correlation between the data and the first-order model.

At a flow velocity of 109 cm/day, TCE, PCE and 1,1,2-TCA fitted reasonably well to the model with r^2

values greater than 0.8. Half lives for TCE, PCE and 1,1,2-TCA were calculated as 1.2, 2.1 and 1.9 hours respectively. Half lives were calculated at 12.5 hours and 5.0 hours for c-DCE/MEK and 1,1-DCE concentrations, giving r^2 values of 0.750 and 0.831. The first-order kinetic model could not be applied to VC and t-DCE due to increasing and variable compound concentrations.

At a flow rate of 54 cm/day, best fit half lives of TCE, PCE and 1,1,2-TCA were 3.7, 5.4 and 3.5 hours respectively, with corresponding r^2 values all greater than 0.945. Half lives of 14 and 24 hours were calculated for c-DCE/MEK with poor correlations of 0.654 and 0.591 respectively. Half lives of 10.6 hours and 8.7 hours were calculated for t-DCE and 1,1-DCE respectively. No 1,1,1-TCA was detected. A half life of 1.9 hours was calculated for TCM based on low concentrations, with an r^2 value of 0.91.

Non-chlorinated breakdown products including methane, ethene and ethane showed reasonably consistent concentrations once steady state was achieved in the columns.

5.2.3.2 Inorganic Contaminants

There was very little evidence of changes in concentration for Na^+, Mg^{2+}, SO_4^{2-} and NO_3^- as site water passed through the column at both flow velocities. Dissolved Fe increased significantly from 0.32 and 0 mg/L to 22.0 and 16.8 mg/L at the faster and slower flow velocities respectively. The concentrations of Mn and K^+ also increased in the effluent. The Cl^- concentration increased due to the dechlorination of TCE and its daughter intermediates. Calcium concentrations decreased between the influent and effluent samples and both alkalinity and oxygen concentrations decreased in the effluent.

The redox potential demonstrated reducing conditions, for high and low flow velocities, with the effluent end of the column for both flow velocities exhibiting only slightly reducing conditions. The pH for both velocities showed a decrease from 7.6 and 7.8 to 5.8 and 6.0.

5.2.4 CONCLUSIONS

TCE degraded very rapidly with a half life of 1.2 to 3.7 hours, generating c-DCE as an intermediate degradation product with the calculated half life of c-DCE/MEK ranging between 12-24 hours. c-DCE was therefore a critical compound in the design of the PRB system.

Calcium concentrations showed little decline, although the loss of alkalinity was substantial. Dissolved iron concentrations increased significantly due to the oxidation by water and the high concentrations of TCE. The decrease in pH values and the loss of alkalinity were attributed to the precipitation of siderite ($FeCO_3$).

The column test demonstrated that a significant plume of dissolved iron would be expected to occur downgradient from the PRB resulting in the precipitation of siderite and iron oxide.

5.3 NUMERICAL MODELLING AND CONCEPTUAL DESIGN

5.3.1 BACKGROUND AND METHODOLOGY

A conceptualisation of the site hydrogeology developed by Golder during the site characterisation programme, was modelled using the two dimensional, finite difference, steady-state groundwater flow model FLOWPATH which simulates groundwater flow in the horizontal plane. The code also contains a particle tracking routine, which plots groundwater pathlines and can be used to demonstrate the capture zone of the reactive barrier.

In a typical model application, a grid is constructed over the area to be studied. Hydraulic boundaries are defined at the edge of the model and hydraulic parameters are assigned to the grid nodes. Hydraulic heads at each grid node are calculated, and the model is calibrated by adjusting the hydraulic parameters until the distribution of calculated hydraulic heads matches the distribution of heads measured in the field.

5.3.2 PURPOSE OF THE MODEL

The purpose of the groundwater flow model was to assist in the design of the PRB system. Specifically, the model was used to:

- Assess whether the PRB system would have any significant effect on the groundwater flow field
- Determine the residence time of groundwater within the reactive barrier
- Determine the required thickness of the reactive barrier
- Determine the alignment and length of the associated cut-off wall

5.3.3 MODEL DESIGN

The modelled area encompassed the eastern half of the site and adjacent land. A rectilinear grid comprising 49 rows and 31 columns was established, with a higher grid density in the immediate area of interest and where there was a greater level of data. The model grid is illustrated in Figure 5.3. The model assumed an unconfined aquifer with an arbitrary base of approximately 11 mbgl. The western boundary of the model was taken as a constant head boundary using the 35 m groundwater head contour. Other constant head boundaries included: 25 m for the northeast boundary; 28 m along the north; and 29 m along the south with a short section set at 31.8 m where flow is towards a small brook beyond Monkstown Avenue. Other boundaries were taken as no flow boundaries.

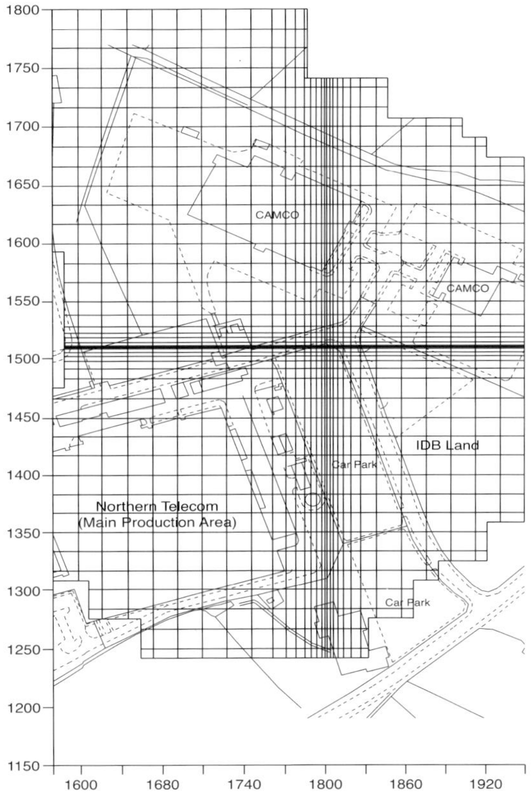

Figure 5.3: Model grid Source: Golder Associates (1994)

The geology of the major portion of the study area was assumed to be till and was assigned two values for hydraulic conductivity: 2.3 x 10^{-7} m/s and 1.2 x 10^{-7} m/s; and a porosity of 0.3. North and east of the car park the aquifer becomes sandy and was assigned hydraulic conductivity values of 1.7 x 10^{-6} m/s and 1.7 x 10^{-5} m/s with a porosity of 0.15. Another zone of high permeability was simulated to the east of the main production area on the site. A low recharge rate of 50 mm/yr was estimated for the site due to site drainage and hardstanding, but areas of enhanced recharge at both the northern end and central portions of the eastern car park area were assigned values of 256 mm/yr and 150 mm/yr respectively.

For the purposes of the initial model run, the reactive barrier wall was given dimensions of 3 m wide by 1 m thick by 10 m deep, a hydraulic conductivity of 5 x 10^{-4} m/s and a porosity of 0.4. The cut-off wall which is typically constructed of bentonite cement grout was assigned a length of 25 m, hydraulic conductivity of 10^{-8} m/s, and a porosity of 0.1. Although the hydraulic conductivity of the reactive barrier is very high, flow through the barrier is controlled by the lower hydraulic conductivity of the aquifer material on either side of it.

5.3.4 MODEL CALIBRATION

The model was calibrated using the water level contour map of May 1994 and water level measurements made in October 1994. A convergence criterion was used that stipulated that the head change at every node after the final iteration would not be greater than 0.05 % of the maximum head difference in the system. The water balance in the simulation closed to within 1.6 %.

5.3.5 RESULTS

The model was run with an initial set of parameter values. Subsequent runs involved changing selected parameter values to determine the sensitivity of the system to those changes.

The hydraulic head distribution based on conditions prior to installation of the PRB was not significantly altered by the construction of the reactive barrier and cut-off wall. The model estimated the head difference across the reactive barrier to be 0.1 m. The average linear groundwater velocity within the reactive barrier was calculated using Darcy's Law to be 1.2 m/d. The residence time within the modelled reactive barrier was estimated to be of the order of one day. Away from the reactive barrier, contaminant movement was estimated to be 5.5 x 10^{-2} m/d or 20 m/a assuming no retardation.

The volumetric flow through the modelled reactive barrier can be determined using Darcy's Law which is described by Equation 5.4:

$$Q = -KiA \qquad \text{Equ. 5.4}$$

where;
Q= volumetric groundwater flow
i= hydraulic gradient = 0.1
K= hydraulic conductivity = 9.9 x10^{-6} m/s
A= cross sectional area = 300m^2

and is calculated to be 30 m^3/d. The value for hydraulic conductivity is an average of the two hydraulic conductivities used to model the high permeability zone in the area where the PRB was to be installed. Whilst the model assumed uniform flow through the reactive barrier, the actual flow rate through the barrier was expected to be considerably less because of the interbedded nature of the aquifer, giving rise to both high and low permeability zones.

Model sensitivity was run on the reactive barrier hydraulic conductivity to determine what would happen in the event that contaminant residence time within the reactive barrier was insufficient to allow full dechlorination of the TCE. Residence time in the reaction cell could be increased by decreasing the hydraulic conductivity of the PRB. Sensitivity analysis showed that if the hydraulic conductivity of the reactive barrier was reduced by an order of magnitude, there would be negligible effect on the hydraulic head drop across the barrier wall, but the average linear groundwater velocity would be reduced to 0.1 m/d and the contaminant residence time within the modelled PRB would be increased to approximately 10 days.

Sensitivity of the system to changes in the length of the cut-off wall was run for the purposes of estimating cost benefits from a reduced length of cut-off wall. It was found that without a cut-off wall, only contaminants in the northeast corner of the eastern car park would pass through the reactive barrier. Adding a cut-off wall and extending it to 10 m in length on either side of the reactive barrier, increased the zone of capture for the PRB, but the model predicted that contaminants distant from the PRB would pass through the cement grout of the cut-off wall rather than be hydraulically captured by the reactive barrier.

Sensitivity of the system to improvements in car park drainage were modelled to assess changes in hydraulic gradient that might arise from reduced recharge. The model was run with reduced recharge of 1.4×10^{-4} m/d. The results indicated that the existing gradient across the reactive barrier would be maintained.

The results of the modelling exercise provided an order of magnitude estimation of system parameters and supported the viability of a PRB system at the site. The model predicted that the hydraulic regime at the site would not be adversely affected by the installation of a PRB system and that contaminants would not be diverted around the cut-off wall.

5.4 DESIGN OF CUT-OFF WALL AND REACTIVE BARRIER SYSTEM

Based on field observations, and modelling, it was decided that the PRB would be placed at the boundary of the eastern car park. Contaminant concentrations at the discharge from the PRB were designed to meet a site specific discharge criterion for TCE of 10 μg/L at a flow rate of 5 m^3/day. Contaminant resident time in the PRB was designed for a minimum of 12 hours.

A cement bentonite cut-off wall would provide a cost effective hydraulic barrier of low permeability to re-direct groundwater toward the reactive granular iron. It was decided to house the granular iron in a vertically aligned steel vessel that could be lowered into an excavation constructed under cement bentonite slurry, which formed part of the cut-off wall. This design addressed a number of concerns, namely:

(i) the source zone was close to a public road. The vertical flow design allowed a greater flow path length in the Fe^0 to be achieved without the PRB encroaching onto the road.

(ii) the source zone materials were heterogeneous in nature. Groundwater flow through the source zone and across an uncontained vertical wall would have been very variable, creating regions of high flow, and thus low residence time within the wall and possibly ineffective treatment. The Monkstown design maintained uniform flow through the reactive cell.

(iii) if not fully effective, an uncontained vertical wall at the site could have conducted contaminants from the upper contaminated subsurface to the lower uncontaminated subsurface. The main flow horizon at the site occurred in an unconfined aquifer. The saturated thickness was expected to be small, requiring a thin but very long reactive cell. Excavation to increase the depth of the reactive cell and thus the depth of flow could have breached an underlying clay aquitard. Using a vertical flow path within the reactive cell, enabled the PRB to be sealed into an enlargement in the cut-off wall with no risk of creating new vertical flow paths in the ground.

(iv) the vertical design allowed for a more controlled monitoring system with which to evaluate the performance of the system.

Plan and section views of the PRB system are provided in Figures 5.4 and 5.5 respectively.

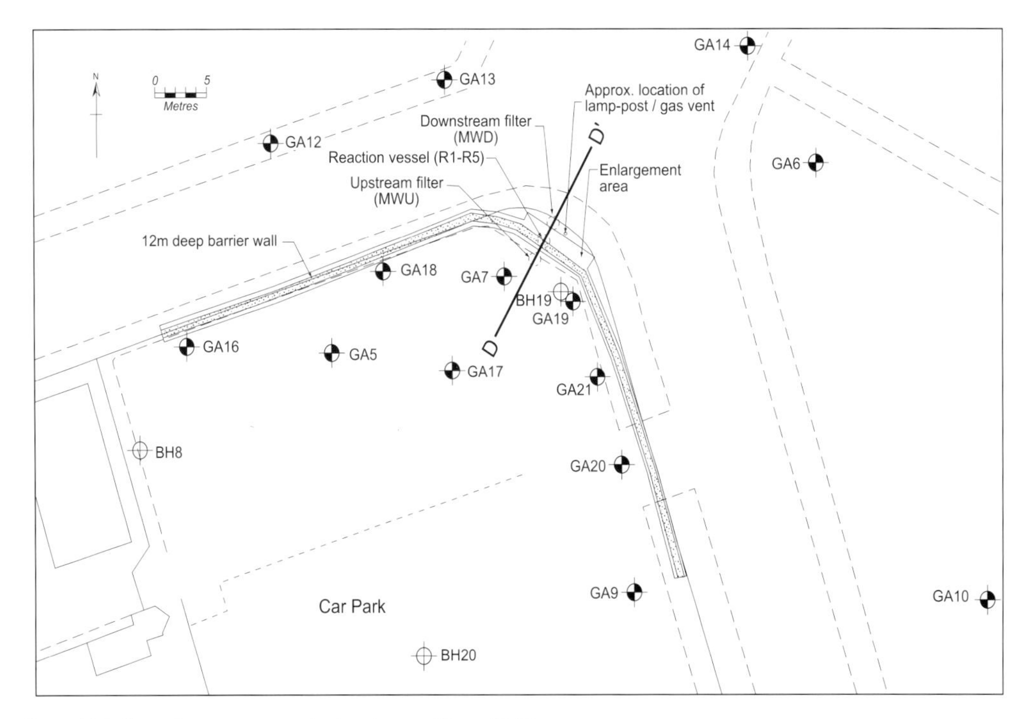

Figure 5.4: Plan of eastern car park showing position of PRB

Source: Golder Associates (1996)

5.4.1 REACTIVE BARRIER

The basic design of the reaction vessel was a 1.2 m diameter steel cylinder 12 m long, divided into three compartments.

The lowest compartment, containing the granular iron was 7.3 m long. This section was fitted with a hinged steel door to maintain the granular iron under anaerobic conditions. This environment was necessary for the dehalogenation process to take place effectively. The rest of the vessel remained empty. A locked steel grate door was fitted within the upper section and a solid, lockable outer hatch, was added to prevent unauthorised entry.

A ventilation pipe was installed to vent the gases to atmosphere and to prevent any build up of gases in either the reaction cell area or the open vessel above. This pipe was fitted into a modified street lamp standard. The lamp standard contained the necessary flame retarders and top of stack dispersal system.

The performance of the vessel is monitored via discrete sampling points, which are accessed by lockable covers outside the vessel to eliminate the need to enter the vessel for sampling. These sampling points can also be used to access and agitate the granular iron should the flow rate decline in the future due to

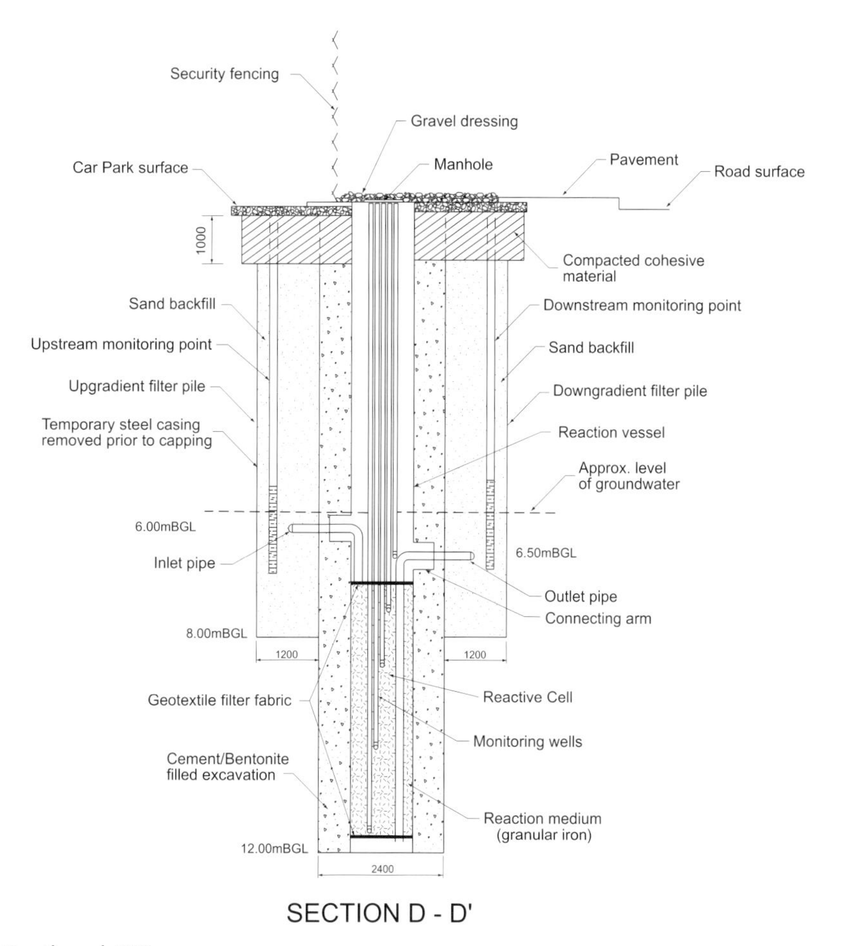

Figure 5.5: Section through PRB Source: Golder Associates (1996)

the build up of mineral precipitates.

The laboratory studies found that the low pH of the system, coupled with the oxidation of the iron by water in the presence of high concentrations of TCE, could precipitate iron carbonate as siderite ($FeCO_3$). The initial conceptual design called for vertical upward counter-current flow through the reactive cell, but due to the concern with potential mineral precipitation and subsequent plugging, at the entrance of the cell, the flow direction was changed to downward. Therefore, any mineral precipitation would take place at the entry (top) of the iron column and could be easily removed.

Granular iron with a grain size ranging from 0.57 mm to 2 mm was sourced in the United States from Master Builders Inc.™ in Cleveland, Ohio, since it was this source of iron which had been used in the bench tests.

6 INSTALLATION OF THE PERMEABLE REACTIVE BARRIER

6.1 INTRODUCTION

This section discusses the aspects relating to the installation of the PRB and includes the following:

- Cut-off Wall Construction
- Reaction Vessel Installation
- Filter Pile Installation
- Auxiliary Pumping

6.2 CUT-OFF WALL CONSTRUCTION

The barrier wall was constructed between November 1995 and February 1996, using a cement bentonite slurry technique. The excavation was dug using a modified, tracked backhoe with an extended boom as illustrated in Plate 6.1.

Plate 6.1: Modified backhoe excavating trench Source: Golder Associates (1996)

The trench was excavated to a depth of 12 m below ground level and a width of 0.6 m. Plate 6.2 illustrates the excavation and backfilling for the barrier wall.

The permeability specification for cut-off walls is a geotechnical specification and has the same units as hydraulic conductivity. The permeability for the cut-off wall at Monkstown was set at 1x 10^{-9} m/s which was achieved in laboratory tests for the cement bentonite after 90 days.

Plate 6.2: Excavated trench Source: Golder Associates (1996)

In preparation, a bentonite slurry was mixed and allowed to hydrate for 24 hours then ordinary portland cement (OPC) and ground granulated blast furnace slag (GGBFS) were added before the mix was pumped into the trench. During construction, the sides of the trench were supported by the cement bentonite slurry which was progressively poured into the trench as it was being cut. The slurry formed a material with a permeability of less than $1x\ 10^{-9}$ m/s. During construction, samples were collected for laboratory testing to confirm that the slurry properties met the requirements of the barrier wall.

The design depth of the wall was set at a maximum of 12 m in order to key the barrier wall into a horizon of low permeability. During excavation some minor modifications in depth were necessary due to the presence of large boulder erratics at 11.5 m below ground level, which were removed.

At the location where the reaction vessel was to be installed, the wall was thickened to accommodate the vessel. This also increased the flow path length through the wall in the source zone and reduced the likelihood of seepage through the wall and around the vessel. This section was also excavated, under cement bentonite slurry, to 12 m below ground level.

On completion of the wall and wall enlargement, the upper 1 m of cement bentonite was excavated and a compacted clay cap was installed on top of the wall. Trimming of the wall to a depth of 1 m ensured that any dried and cracked cement bentonite material was removed from the top of the wall and that a sufficient depth of clay capping could be achieved, thereby removing potential preferential pathways for shallow groundwater flow through the wall. The clay capping was also extended over the filter piles (see Section 6.4) to prevent surface water within the hardcore fill of the car park from entering the system.

6.3 REACTION VESSEL INSTALLATION

The reaction vessel was installed between December 1995 and January 1996 in the thickened section of the cement bentonite wall as shown on Figure 5.4. Due to the rapid curing time of the cement bentonite slurry it was necessary to complete the construction of the thickened section and installation of the reaction vessel in one phase.

Once the excavation had been prepared, the reaction vessel was lifted and sufficient granular iron was placed in the lower compartment to ensure that it would sink in the slurry. The filling process generated iron dust which necessitated the use of dust control measures to ensure safe working conditions. The reaction vessel was then lowered into the excavation. Plate 6.3 illustrates the enlarged slurry wall with the vessel lying horizontally in the background. Plate 6.4 shows the reaction vessel vertically aligned and suspended above the slurry wall with the flow connection arms near the base of the vessel.

Plate 6.3: Wall enlargement for reaction vessel. Reaction vessel in the background. Source: Golder Associates (1996)

Due to the physical properties of the slurry, the setting time required to allow sufficient bearing capacity to support the filled reaction vessel was considered excessive. To overcome this problem, neat cement slurry was pumped to the base of the excavation to provide additional end bearing support. Once the reaction vessel was in the correct position and met tolerances for vertical alignment and position, it was secured in place and the slurry was left to set. Once the slurry and neat cement grout had hardened to sufficient strength, the remainder of the granular iron was added.

6.4 FILTER PILE INSTALLATION

In order to create high permeability catchment zones for groundwater at the entrance and exit of the reaction vessel, filter piles were constructed on the upgradient and downgradient sides of the wall (see Figure 5.4 for detail). The piles were auger bored to a depth of 8 mbgl. The lower 2 m was backfilled with a clean, well-graded sand and the casing was partially removed. At this point the cement bentonite

Plate 6.4: Reaction vessel aligned for installation Source: Golder Associates (1996)

in the wall enlargement had hardened sufficiently to allow the two connecting arms to be excavated out towards the filter piles. The excavation was carried out from within the reaction vessel shell. Using this technique, it was not necessary for anyone to enter the potentially solvent-rich atmosphere of the filter piles. The water level in the filter piles was maintained at a low level during this period by pumping. Once the arms were connected, the casing was removed from the piles and they were filled to the surface with clean sand.

The use of a drainage trench excavated under polymer slurry and backfilled with gravel was considered. However, at that time, the compatibility of polymer residues with the iron had not been investigated.

6.5 AUXILIARY PUMPING

Monitoring well GA13, which is located outside the boundaries of the site and downgradient of the PRB system (see Figure 5.4), contained high concentrations of TCE in both soil and groundwater. In order to capture the contaminant plume moving away from GA13, an auxiliary pumping system, both solar and wind-powered, was constructed to pump contaminated groundwater from GA13 and recharge it through monitoring well MWU, located directly upgradient of the PRB (see Figure 5.4). The auxiliary pumping system was taken out of commission in December 1999 when it suffered wind damage, and was reinstated in February 2001.

7 PERFORMANCE MONITORING

7.1 INTRODUCTION

Following the installation of the PRB, a groundwater monitoring programme was established to verify whether the system was operating as designed. The monitoring programme consisted of water level readings and geochemical sampling. Water levels were measured to ensure that the PRB system had not adversely affected the groundwater conditions. Geochemical sampling of groundwater upgradient, within, and downgradient of the reaction vessel was conducted to provide a means of assessing actual changes in groundwater chemistry of the full scale PRB system against expected results predicted from the laboratory scale column tests, and to demonstrate that discharge from the reactive cell meets design criteria.

Following the installation of the PRB, groundwater sampling stations were established at the following locations: wells upgradient of the reactive cell: GA5, GA7, GA17, GA19, GA21, BH19, and MWU; sample stations within the reactive cell: R5 (entrance), R4, R3, R2 and R1 (exit), and wells located downgradient of the reactive cell: GA6, GA10, GA12, GA13, GA14 and MWD. A layout of the sampling stations is provided in Figure 7.1.

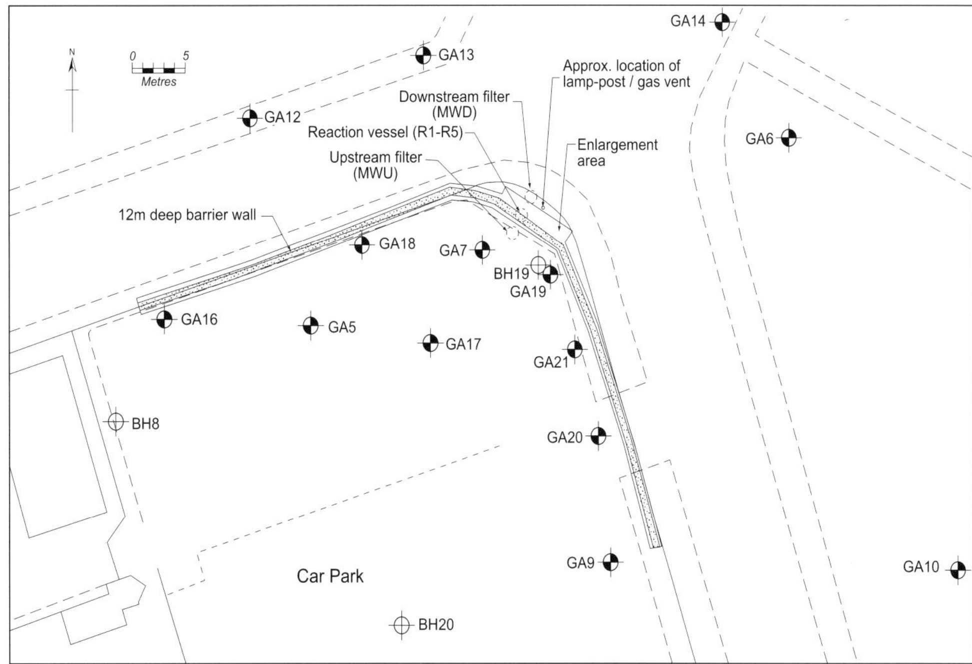

Figure 7.1: Layout of additional sampling stations Source: Golder Associates (1996)

7.2 WATER LEVEL MONITORING

Groundwater levels were measured at least weekly from 20 November 1995 until April 1996, after which time the water levels were measured monthly until October 1996 and then twice yearly until January 1999. Groundwater samples were collected quarterly in 1996, twice yearly in 1997 and annually in 1998,

1999, 2000 and 2001. Groundwater levels and groundwater samples were measured and collected in accordance with protocols developed for the site (see Section 4.7). A summary of groundwater levels is provided in Table 7.1 from a selection of the boreholes.

Table 7.1: Summary of groundwater levels

BH No.	BH elevation (maOD)	21/11/95	1/3/96	18/6/96	20/1/97	11/8/97	3/4/98	16/12/98	28/1/99
GA7	36.07	30.65	32.72	32.17	32.34	30.89	32.67	33.07	33.31
GA17	36.07	30.60	32.86	32.07	32.25	31.95	32.87	33.28	33.37
BH8	36.32	30.36	31.28	n/r	30.99	30.67	30.24	35.29	35.31
GA6	36.75	29.64	30.01	29.89	29.97	29.66	30.02	30.02	n/r
GA14	36.15	30.58	31.85	31.18	31.33	36.15	31.13	31.28	32.25

n/r = no reading
maOD = metres above Ordnance Datum

Source: QUB (2001)

Water level monitoring confirms that generally groundwater flow is north to northeasterly. Gradients range from 0.016 to 0.11 between wells located on either side of the PRB.

Water levels in MWU and MWD indicated generally low and variable gradients within the reaction vessel, with indications of gradient reversals and thus reversals of groundwater flow across the reaction vessel. Based on the existing data it is not possible to accurately calculate the amount of groundwater flowing through the PRB.

7.3 GROUNDWATER CHEMISTRY

A summary of major ion chemistry and selected volatile organic compounds including: the main toxicity drivers, TCE and its degradation products c-DCE and VC; and TCA is provided in Appendix 2. Groundwater chemistry is discussed below.

7.3.1 UPGRADIENT SAMPLING POINTS

Upgradient monitoring wells, BH19, GA7, GA19 and GA21 were sampled every 3 to 6 months from 1994 until 1998 and annually thereafter. Monitoring well MWU has been sampled since 1996, and monitoring wells GA5 and GA17 have been sampled annually since 1998.

Major ion chemistry in the upgradient wells indicates that the major water type is 'calcium bicarbonate' with lesser but significant levels of sodium and magnesium as illustrated on a Piper diagram in Figure 7.2. The data is generally consistent with the background chemistry for upgradient wells proximal to the PRB sampled prior to its installation as described in Section 3.4.

Chemical analyses show that TCE is the major contaminant in upgradient wells with concentrations ranging from a maximum of approximately 390,000 µg/L which was reported in GA19 in August 1994 to a minimum of 4 µg/L in GA5 which was reported in January 1999. See Figure 7.3 for details. Three wells, BH19 in March 1994 and GA19 and GA21 in April 1994 reported 250,000 µg/L of TCE. Lesser amounts of c-DCE, ranging in concentration from below detection to 6200 µg/L, were reported. VC was detected on only one occasion in well MWU in April 1998 at a concentration of 530 µg/L.

Most monitoring wells upgradient of the PRB showed a trend toward decreasing concentrations of TCE over time. For example, TCE in BH19 was initially reported at 250,000 µg/L in March 1994, and steadily

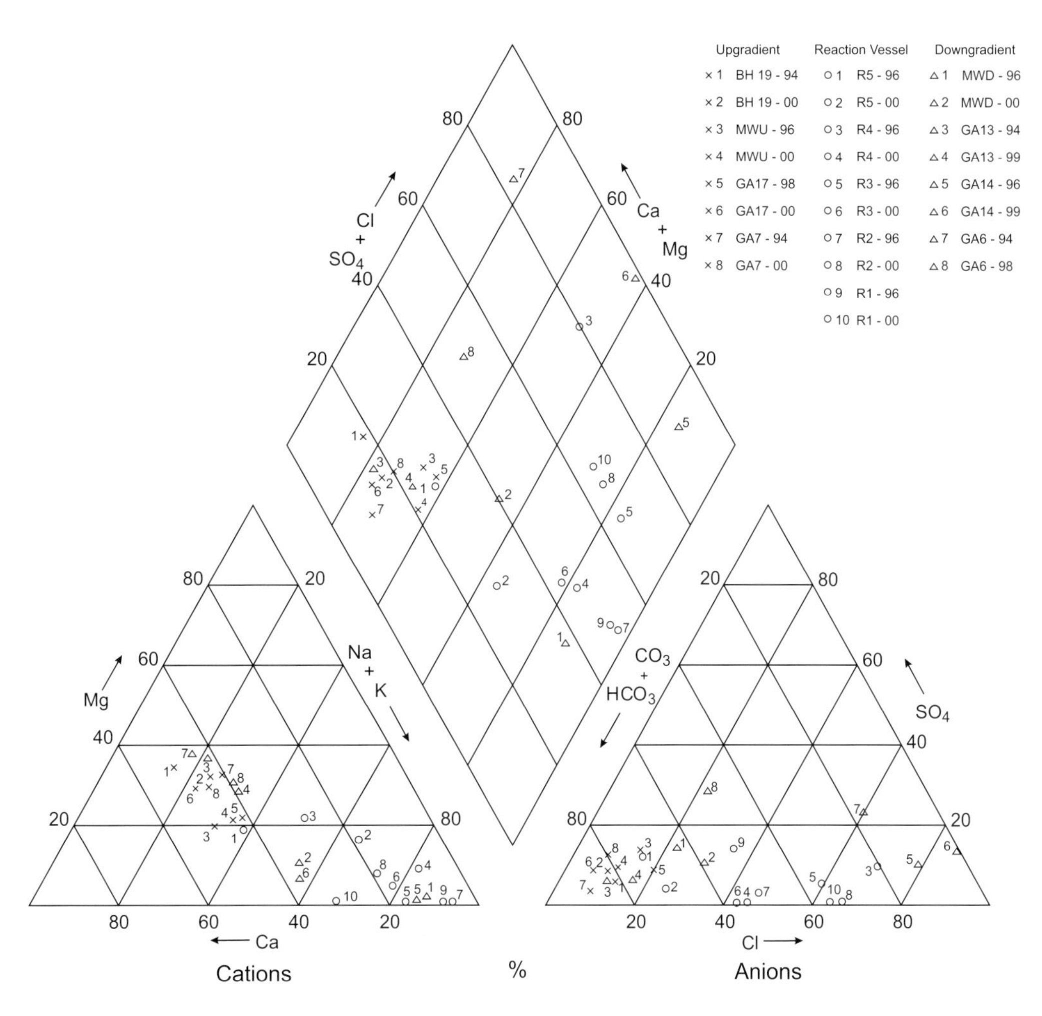

Figure 7.2: Major ion chemistry illustrated on a Piper diagram

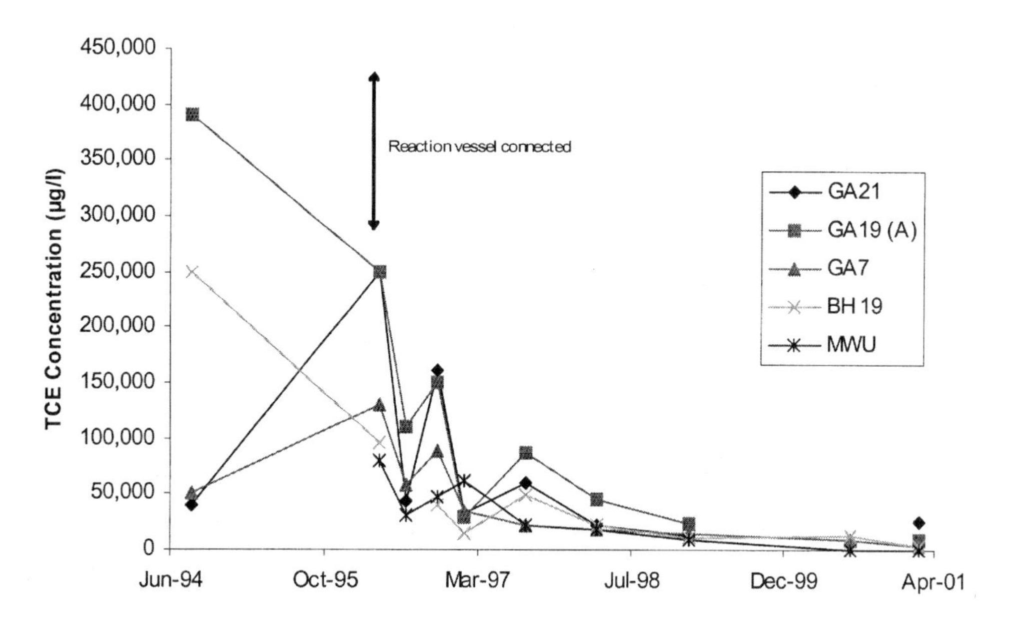

Figure 7.3: TCE concentrations in upgradient monitoring wells

Source: Golder Associates (2001)

decreased to around 4,000 µg/L in 2001. An exception to the trend occurs in GA17 which reported 360 µg/L in April 1998 and increased to 730 µg/L in July 2000 and then decreased to 50 µg/L in February 2001. Generally, the lower concentrations of TCE remain significantly higher than the design criteria of 10 µg/L. The reason for the rapid decrease in TCE concentration in some of the upgradient wells closest to the PRB may be explained by the removal of highly contaminated aquifer material from the trench during installation of the PRB and cut-off wall, although some contaminated material remains; and/or the tail end of a slug of contamination that moved through the site. The degree to which natural variation, natural attenuation, seasonal fluctuation and disturbance during drilling/excavation have affected particularly those wells with lower concentrations of TCE cannot be determined from the existing data.

7.3.2 REACTION VESSEL SAMPLING POINTS

Reaction vessel sampling points (MWU, MWD, R1, R2, R3, R4, R5) have been monitored regularly since their installation in 1996. Major ion chemistry indicates that groundwater enters the reaction vessel as 'sodium-calcium bicarbonate' type and leaves the reactive cell with a slight increase in chloride and less bicarbonate. This supports laboratory column experiments and the mineralogical observations that calcium carbonate (calcite) is being precipitated within the reactive cell (see Section 8.5.2.1).

The contaminant chemistry shows maximum concentrations of TCE and c-DCE in sampling point R5, which is located at the upgradient entrance to the reactive cell. See Figure 7.4 for details. TCE in R5 ranged from a maximum concentration of 38,000 µg/L in April 1996 to 450 µg/L in February 2001, reflecting the trend of decreasing concentrations over time shown in the upgradient wells. c-DCE in R5 ranged in concentration from 57 µg/L in April 1996 to 7200 µg/L in August 1997 before dropping to 57 µg/L in February 2001. VC was not detected in R5. Contaminant concentrations show successive decreases from R4 through to the exit at R1. VC was detected on three occasions in R4, at concentrations of 0.4 µg/L in April 1996 and 2 µg/L in August 1997 and April 1998 respectively. TCE in R1 ranged from a maximum of 25 µg/L in April 1996 to below the limit of detection in February 2001.

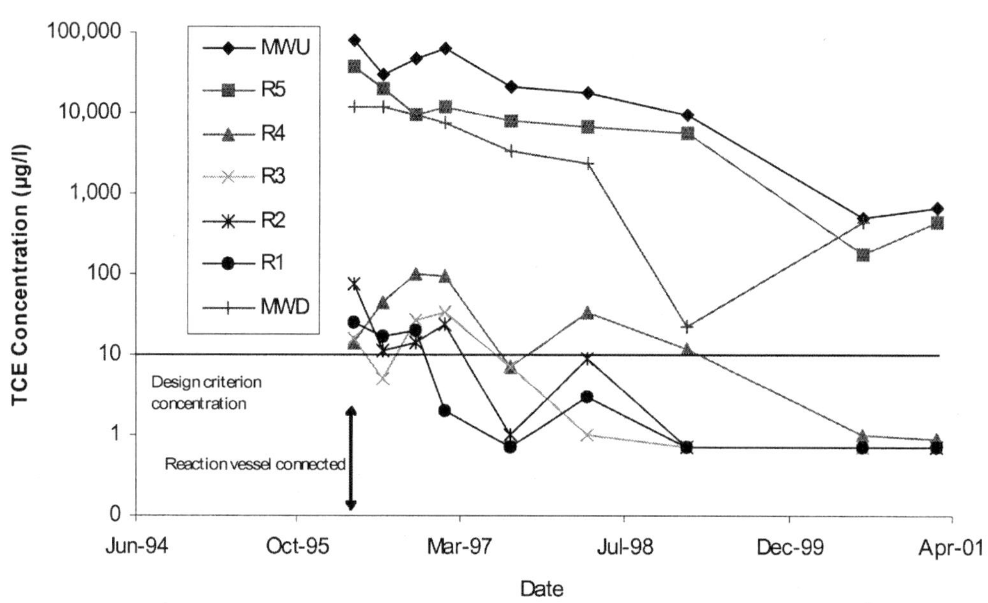

Note: TCE concentrations recorded as not detected have been assigned a value of 0.7µg/L which equals the lowest reported value by the laboratory.

Source: Golder Associates (2001)

Figure 7.4: TCE concentrations in reaction vessel monitoring wells

7.3.3 DOWNGRADIENT SAMPLING POINTS

Sampling points GA12 and GA13 have been regularly monitored since 1994. GA6 has been sampled intermittently since 1994. GA10 was monitored from 1994 until 1997. GA14 and MWD have been monitored since 1996. See Figure 7.5 for details.

Major ion chemistry in the downgradient wells show a range of different water types, from bicarbonate-rich to chloride-rich with varying amounts of calcium, magnesium and sodium as illustrated on the trilinear plot in Figure 7.2. The highest levels of sodium and chloride may reflect sources other than the precipitation of calcium carbonate such as release from the bentonite wall and/or some input from road salting.

Contaminant chemistry shows potential impact of the PRB in wells closest to the reactive cell with little or no impact on more distant wells. MWD, which is closest to the reactive cell, shows a decrease in concentration of TCE from 12,000 µg/L in March and July 1996 to 440 µg/L in July 2000. The elevated levels of TCE in MWD following installation of the PRB reflect the generally low hydraulic gradients and possibly residual levels of TCE in the superficial deposits downgradient of the PRB. The decrease in downgradient contaminant concentrations reflects a similar trend in the upgradient wells. GA6, GA10, and GA12 also show a decreasing trend in TCE. GA13 shows increasing TCE concentrations from 1994 to 1996 with a maximum concentration of 260,000 µg/L in March 1996. Thereafter, the concentrations show a declining trend to approximately 37,500 µg/L until 1999. In February 2001 TCE concentrations increased to 61,000 µg/L indicating that residual contamination is still likely to be present in the vicinity of GA13. GA14 shows low contaminant concentrations compared to other wells since the installation of the PRB, with TCE concentration ranging from below detection to a maximum of 2 µg/L. This suggests that GA14 was located in an area that did not contain historic contamination. c-DCE was detected in most wells with a maximum concentration of 9500 µg/L in GA6 in August 1997. VC was not detected in any of the downgradient wells.

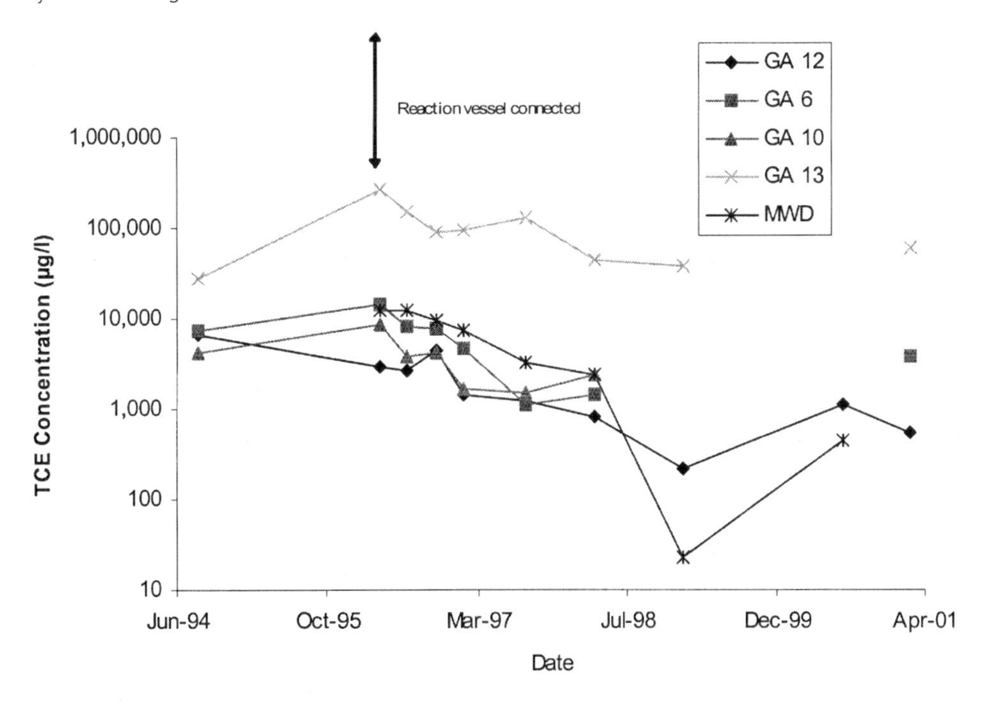

Source: Golder Associates (2001)

Figure 7.5: TCE concentrations in downgradient monitoring wells

7.3.4 SUMMARY

Monitoring of water levels within the reaction vessel itself indicates periodic reversals of groundwater flow across the reactive cell, making groundwater flow through the reaction vessel difficult to quantify. Groundwater flow through the reaction vessel can be estimated using hydraulic parameters derived from other areas of the site.

The major ion chemistry shows the predominant groundwater type upgradient of the PRB to be 'calcium bicarbonate'. Groundwater passing through the reactive cell changes from 'calcium bicarbonate' type to 'magnesium-sodium sulphate-chloride' type indicating loss of calcium bicarbonate due to calcite precipitation.

Significant decreases in TCE concentrations in some upgradient wells can be explained by the removal of highly contaminated material during excavation of the PRB and cut-off wall, but the degree to which natural variation, natural attenuation, seasonal fluctuations and disturbance during drilling/excavation affects TCE concentrations in wells cannot be determined from the existing data.

The reactive cell appears to be operating effectively based on contaminant concentrations exiting the cell at R1. Concentrations of TCE have been progressively removed as groundwater flows down through the reactive cell, and the discharge from the reactive cell meets the design criterion concentration of 10µg/L TCE.

8 PERFORMANCE EVALUATION

8.1 INTRODUCTION

Performance monitoring and evaluation of PRBs was discussed generally in Section 2.5. PRB performance is related to the design criteria and the clean up criteria that have been established for the site. The performance of the Monkstown PRB is discussed in this section with attention given to the following topics:

- effective plume capture
- sufficient residence time in the reactive cell
- initial reactive cell performance, and
- long term reactive cell performance.

8.2 PLUME CAPTURE

Verification that the wall and collection system provides effective capture of the contaminated groundwater requires demonstrating that the cut-off wall system is not being bypassed either laterally or vertically by groundwater flow.

Groundwater flow around or beneath the cut-off wall can be demonstrated by measuring contaminant concentrations in strategically placed monitoring wells in areas of potential plume transgression. The cut-off wall at Monkstown is keyed into an aquitard at 12 mbgl and significant flow underneath the wall is not considered likely. Strategically placed wells on the near downgradient side of the cut-off wall include GA 10, GA 6, and GA 12. All of these wells show residual but decreasing levels of TCE from historical sources which existed prior to the installation of the PRB system. Until the concentration of TCE decreases and stabilises to levels significantly below concentrations upgradient of the PRB, downgradient wells cannot be used to monitor plume transgression at Monkstown.

8.3 RESIDENCE TIME

The ability of the reactive cell to provide complete dechlorination of contaminants in groundwater entering the reactive cell is dependant on the groundwater having sufficient residence time within the reaction vessel. Residence time can be calculated directly by means of tracer tests, or by modelling groundwater flow through the cell. That the TCE has experienced sufficient residence time in the reactive cell can also be determined by monitoring the concentrations of TCE and degradation products discharging from the cell.

8.3.1 TRACER TEST

A tracer test was carried out within the reactive cell, which involved the injection of bromide as a conservative tracer in MWU, but the test was unsuccessful because of suspected sorption of the tracer, and no further testing has been carried out to date (D. Haigh, Golder Associates - personal communication).

8.3.2 GROUNDWATER FLOW

Residence time can be calculated indirectly by estimating system parameters and applying Darcy's Law.

The reversals in hydraulic gradient observed in the MWU and MWD make it difficult to interpret and model groundwater flow through the reaction vessel. However, estimates of flow can be made by using hydraulic parameters derived from other areas of the site. If it is assumed that the cut-off wall is capturing all of the groundwater entering the capture zone, then the volume of captured water is a good approximation of the maximum flow through the reaction vessel. The volumetric groundwater flow can be estimated from Darcy's Law:

$$Q = -KiA \qquad \text{Eqn. 8.1}$$

(see definition of terms Equation 5.4)

Due to the paucity of field measured hydraulic conductivity values the hydraulic conductivity, K, was estimated from the original values that were inserted in the model described in Section 5.3.3. Values of 1.7×10^{-5} m/s and 1.7×10^{-6} m/s were selected for silty sands and gravels, while 1.2×10^{-7} m/s and 2.3×10^{-7} m/s were selected for silty, clay tills. The proportion of the cross sectional area which comprises sands and gravel versus clay till was estimated to be 50 % of each based on the cross sections through boreholes GA7 and GA11 located in the vicinity of the reaction vessel.

The regional gradient can be estimated from measuring the difference in head between adjacent boreholes divided by their horizontal separation. Gradients were calculated between the following three sets of wells based on water level measurements taken on more than 40 different occasions between 1995 and 1999 and averaged: GA17 and GA10; BH8 and GA13; and BH20 and GA10. The average calculated gradients for the three zones were: 0.05; 0.016; and 0.11 respectively. The gradient can also be estimated from the piezometric surface map in Figure 3.4 which gives gradients across the eastern car park area ranging from 0.054 to 0.069. For this calculation a gradient of 0.05 was used.

The cross sectional area or capture area for the cut-off wall can be calculated from the saturated depth which was taken as 10 m, and the horizontal distance across the open ends of the wall which was calculated at 52 m, providing a cross sectional area of 520 m^2.

On the basis of these estimates, the volumetric flow ranges from approximately 1 to 6 m^3/d, giving a residence time in the reactive cell of between 17.4 and 105 hours.

8.3.3 MONITORING

Sufficient residence time within the reactive cell can be inferred if the concentrations of TCE in the effluent stream of the reactive cell, meets the design criteria or is below detectable levels. R1 has been selected as representing the effluent stream of the reactive cell. Monitoring results are provided in Appendix 2. Concentrations of TCE ranged between 25 µg/L in 1996 to below detectable levels of 0.7 µg/L . c-DCE results ranged between 2 µg/L in 1996 to below detectable levels since 1999 and VC results have been below detectable levels since 1996. The monitoring data show that with the exception of April, July and October 1996, shortly after installation, in which TCE levels were 25 µg/L, 17 µg/L and 20 µg/L respectively, the reactive cell has met the design criterion concentration of 10 µg/L of TCE for the past 5 years.

8.4 EARLY REACTIVE CELL PERFORMANCE

It is useful to compare observed changes in water quality through the reactive cell against changes predicted on the basis of the earlier laboratory column experiments. These results can assist in understanding the geochemical processes operating at the site, and the potential impact to groundwater quality downgradient of the PRB.

Reactions between influent groundwater and the Fe^0 in the reactive cell will result in changes to the groundwater quality discharging from the cell. Laboratory column experiments predicted degradation of TCE to non-detectable levels. Levels of K^+, Cl^-, dissolved iron and manganese were predicted to increase through the reactive cell, while Ca^{2+}, alkalinity, oxygen and pH were predicted to decrease. No changes were predicted for Na^+, Mg^{2+}, SO_4^{2-} and NO_3^- ; however, these parameters would be expected to be quickly restored to local background values on exit from the reactive cell due to interaction with the aquifer geology and mixing with groundwater.

Table 8.1 shows the concentrations of TCE, c-DCE and selected inorganic analytes measured in MWU, R5, R4, R3, R2, R1 and MWD for an early time period following the installation of the reactive cell, and approximately three years after installation. These data are plotted as concentrations versus monitoring well locations on Figure 8.1.

Table 8.1: Ananlytes concentrations versus monitor well locations

Date	Location	TCE (µg/L)	c-DCE (µg/L)	Chloride (mg/L)	Sulphate (mg/L)	Alkalinity (mg/L)	Ca (mg/L)	Fe (mg/L)	pH
July 96	MWU	30000	270	37	50	220	21.8	1.11	7.55
Jan 99	MWU	9630	1454	30.5	37	138	37.9	3.86	7.6
July 96	R5	20000	420	51	13	230	42.1	0.13	8.19
Jan 99	R5	5564	459	38.5	40	146	37.6	6.71	8
July 96	R4	44	8	104	11	60	23.4	3.13	8.71
Jan 99	R4	12	362	47.5	<0.1	44	6.42	2.1	8.9
July 96	R3	5	<0.3	109	8	50	62.1	0.68	9
Jan 99	R3	<0.7	<0.3	87	<0.1	36	19.6	1.52	8.9
July 96	R2	11	0.3	116	3	50	29.3	1.1	9.58
Jan 99	R2	<0.7	<0.3	68	<0.1	42	18.2	2.35	9
July 96	R1	17	0.5	46	22	110	6.93	0.09	9.67
Jan 99	R1	<0.7	<0.3	46.5	<0.1	46	16.1	2.51	9.8
July 96	MWD	12000	430	85	123	180	5.97	0.3	10.59
Jan 99	MWD	23	4099	288	17	86.1	73.3	3.24	7.9

Source: Golder Associates (2001)

Some of the data show that the reactive cell is generally operating as expected based on results from the laboratory column experiments. Concentrations of TCE and c-DCE are reduced to below detection within the cell at R1 indicating that reductive dehalogenation is occurring. Alkalinity shows a significant drop from R5 to R4 at the entry of the reactive cell while calcium shows a rather "spiky" decline from R5 to

R1 suggesting that precipitation of calcium carbonate is occurring within the reactive cell as predicted. Other data are ambiguous. Iron and chloride show variable trends opposed to the predicted increase, while pH shows a slight increase compared to a predicted decrease. The variable trends may be attributed to variations in flow through the reactor as evidenced by gradient reversals from field measurements.

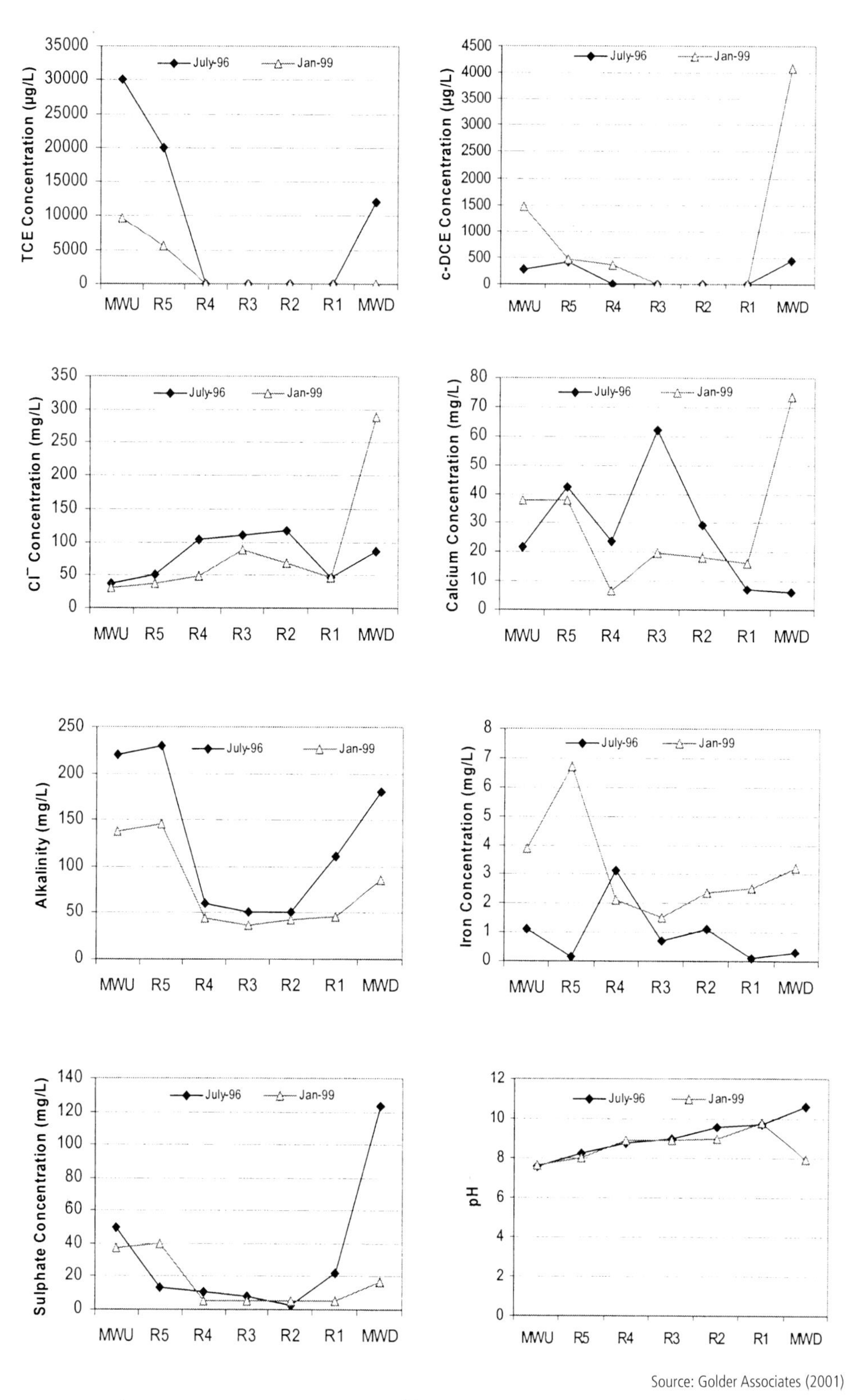

Source: Golder Associates (2001)

Figure 8.1: Analyte concentrations versus monitor well locations

8.5 LONG TERM REACTIVE CELL PERFORMANCE

Changes in the infrastructure associated with the reactive cell with time are required to assess the long term performance of the system. Infrastructure includes:

- the cut-off wall
- reaction vessel, and
- the granular iron within the reactive cell

8.5.1 THE CUT-OFF WALL AND REACTION VESSEL

The integrity of the cut-off wall is controlled by methods of construction and installation, the reaction of groundwater, and man-made interference following installation. The wall was installed in accordance with a draft report which provides specifications for the construction of slurry trench cut-off walls and barriers (ICE *et* al, 1999). Walls constructed to that specification and under the conditions found at Monkstown could be expected to have a design life of 100 years (Professor S. Jefferis, University of Surrey - personal communication). Significant deterioration of the cut-off wall is not expected in the near future based on existing groundwater chemistry. Detailed monitoring of effects on walls in other more aggressive groundwater conditions in the UK are being undertaken, and will provide useful information on possible impacts to the cut-off wall at Monkstown. Man-made interferences from future excavation will be mitigated by ensuring that the location of the PRB is known and appropriately identified.

The operating life of the reaction vessel is estimated to be 50 years, based on wall thickness and estimated corrosion rates (Mr R. Essler, Keller Ground Engineering Ltd - personal communication).

8.5.2 GRANULAR IRON WITHIN THE REACTION VESSEL

As Fe^0 reacts with groundwater, minerals will be expected to precipitate on the surface of the iron resulting in a decrease in the surface area of Fe^0 that is available to react with TCE. Continued build up of precipitates over time will cause a decrease in the rate of contaminant degradation due to decreased surface area, and could cause a decrease in the permeability of the Fe^0 in the reactive cell due to increased volume of solids and loss of void space. Recent work by Gillham (2001), shows that the precipitation of carbonate on the surface of the iron reduces the iron activity, causing the zone of active carbonate deposition to move slowly into the reactive cell bed. If the iron is reasonably coarse, the deposition will cause minimal change in permeability. However, it will reduce the effective bed length, resulting in a decrease in the distance travelled by the water, and a reduction in the residence time. Allowance for this should be made at the design stage. In the case of Monkstown, the PRB was designed to allow access to the granular iron for the purpose of carrying out maintenance and replacement.

Work was carried out by Queens University Belfast to assess the performance of the Fe^0 in the Monkstown PRB. This work included a mineralogical study, assessment of biological activity, calculation of decay rates, and carbon isotope fractionation studies of reactive cell core material. The work is described and discussed below.

8.5.2.1 Study of Reactive Cell Cores

Core collection

In July 1999, the reaction vessel was opened and two samples of granular iron were collected. Initial inspection of the granular iron showed that a crust had formed on the iron on the top surface of the reactive cell at the groundwater entry point. The crust was broken and found to be approximately 5 cm thick. The crust material was brittle, with significant black precipitates throughout. A greenish layer, 2-4 mm thick was observed within the crust and was believed to be the remnants of the geotextile layer placed on the surface of the granular iron after initial iron filling in 1996. Visually, the geotextile layer appeared not to have retained any structural integrity. No evidence was found of any decrease in permeability due to the presence of the crust. Pieces of crust approximately 3-10 cm wide and 2-3 cm thick were collected for analysis.

Samples below the crust were taken from approximately 15-40 cm into the iron. Observation of this iron showed a large amount of black precipitate thought to be iron sulphides. The groundwater brought up with the iron samples was also black, containing fine suspended particles.

Groundwater samples from the reactive cell were collected by Golder and QUB using dedicated check valve pumps within each of the five reactive cell monitoring wells (R1-R5). Samples were collected for VOC and stable carbon isotope analysis in appropriate, laboratory prepared bottles.

During the study, a sample of the original "unused" granular iron was supplied by EnviroMetal Technology for use as a control.

Mineralogical study

Granular iron samples were studied using scanning electron microscopy (SEM) at magnifications of 50X, 300X and 3000X. Images are provided in Figures 8.2, 8.3 and 8.4 at 300X. The control sample shows a distinct grain size pattern (Figure 8.2). The image of the entrance sample shows evidence of growth of precipitates within the first 20 cm of the iron zone (Figure 8.3). The precipitates were predominantly siderite ($FeCO_3$) and calcite ($CaCO_3$), (Figure 8.4) identified by means of proton induced x-ray emission spectroscopy and were predicted by the earlier laboratory column experiments. Precipitates are thought to have formed from the conversion of bicarbonate to carbonate due to the increase in pH resulting from corrosion of the iron (Schüth *et al.*, 2000). Core samples taken closer to the centre of the reactive cell showed minimal formation of precipitates but demonstrated flaking of the iron surface due to corrosion.

Biological activity

Evidence of biological activity can be obtained by direct observation or inferred from stable carbon isotope studies. No significant colonisation of the zero valent iron by micro-organisms was observed by SEM. Stable carbon isotopic fractionation studies of the Monkstown Fe^0 described by Schüth *et al.*, (2000) demonstrated isotope fractionation values within the range established for abiotic degradation by zero valent iron as reported by Bill *et al.*, (2000). The results showed no evidence for significant biological activity within the iron.

Figure 8.2: Electron microscopy image of unused iron Source: QUB (2001)

Figure 8.3: Electron microscopy image of iron within the reactive cell Source: QUB (2001)

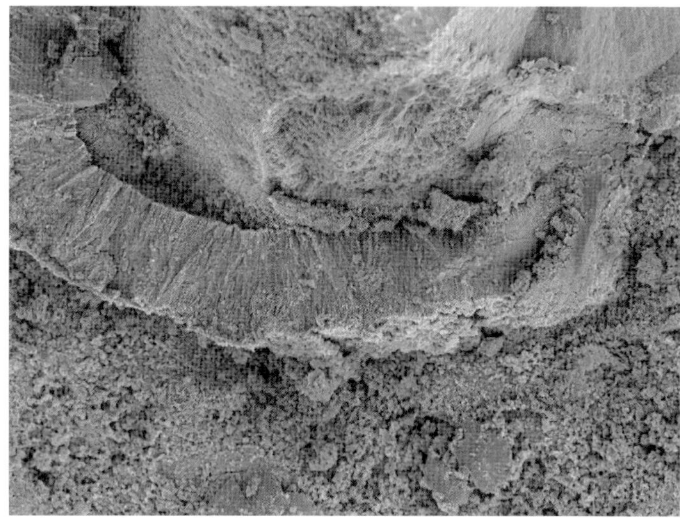

Figure 8.4: Electron microscopy image of the iron crust Source: QUB (2001)

8.5.2.2 Degradation Rates

Batch experiments were set up to measure the degradation rate of TCE using granular iron from the reactive cell "entrance (crust)" and "centre" and from the "control" sample as described in Schüth *et al.*, (2000). The results provided a pseudo first-order reaction rate.

The rate constant, k_l,for the unreacted or control sample was found to be almost identical to feasibility study column experiments carried out by University of Waterloo in 1995 on the granular iron (corrected for differences in experimental design) (Dr K. Walsh, QUB - personal communication). These results suggest that the activity of the "control" iron has not significantly changed as a result of exposure to air during the past five years. The k_l for the entrance sample showed a decrease in value compared to the control sample. This is probably due to a decrease in activity resulting from the build up of siderite and calcite precipitates on the granular iron. The centre sample exhibited a 2-fold increase in the calculated value of k_l, attributable to the increase in surface area (control = 4.2 m^2/g, centre = 11.4 m^2/g) and reactivity from corrosion of the granular iron. This increase in activity of the middle zone more than offsets a small loss of activity in the entry zone.

8.6 SUMMARY

Monitoring wells downgradient of the PRB exhibit detectable concentrations of TCE and DCE as a result of residual historic contamination still present in the subsurface. Until such time as TCE levels stabilise and reduce to levels below those found in upgradient wells, monitoring wells cannot be used to confirm capture of the contaminant plume because their concentrations could mask transgressions of the plume through the cut-off wall.

Groundwater flow through the reactive cell based on potential capture by the cut-off wall is estimated at 1-6 m^3/day, with residence time ranging from 17.4 to 105 hours.

Estimates of volumetric flow and residence time along with non-detectable concentrations of TCE and DCE in groundwater leaving the reactor at R1 confirm that the PRB is operating as designed and meeting the design concentration criterion of 10 µg/L TCE.

The findings from the QUB work show that there was some loss in the activity of the granular iron at the entrance to the Monkstown PRB, but 10-40 cm into the reactive cell centre, the iron activity had improved due to corrosion and increased surface area. The loss of activity in the entry zone is more than offset by the increase in activity in the middle zone.

Mineralogical observation showed the presence of precipitates in the iron sample at the entry point of the reactive cell. After a significantly longer time-period, it is expected that precipitates will cause a decrease in the permeability of this section of the reactive cell. The studies showed that the extent of the precipitation is limited to a zone at the entrance to the reactive cell approximately 5-10 cm in thickness.

Any significant reduction in reactive cell permeability by precipitation of siderite and calcite leading to restricted flow through the reactive cell, would result in an increase in water levels in upgradient wells. The mineralogical evidence shows that the precipitation reactions occurred over a very narrow zone at the entrance to the reactive cell. Permeability and groundwater flow could be restored by excavating and replacing the "encrusted" iron or by high pressure flushing of the affected area.

Based on the field evidence the minimum life expectancy of the iron before any form of maintenance might be required is 10-15 years.

No evidence for significant biological fouling was observed.

9 ECONOMIC CONSIDERATIONS

9.1 INTRODUCTION

Economic considerations are based on the various costs associated with both the overall site characterisation and the PRB remediation programme, including ongoing monitoring. A summary of costs is given in Table 9.1. A cost for alternative options including excavation and landfill disposal with pump and treat, and containment with pump and treat is provided in Section 9.5. A summary table of costs is provided in Table 9.2. Discussions relating to the cost effectiveness of the PRB system are included in Section 9.6.

Table 9.1: Summary of costs of the permeable reactive barrier

DESCRIPTION OF MAIN ACTIVITES	FINAL COST £
Site Investigation	
-Main Site Investigation	192,524
-Additional Site Investigation	37,016
Sub total	**229,540**
Remediation	
- Soil removal and disposal costs	75,000
- Pilot Scale Evaluation	18,000
- Design, Preparation of contracts & working plan	16,000
- Installation of cut-off wall and PRB	252,260
- Supervision	38,041
- Completion Report	10,510
Sub Total	**409,811**
Groundwater Monitoring	
- Monitoring (10 years)	88,193
- Tracer Test	8,000
- Consumables	200
Sub Total	**96,393**
TOTAL COSTS (£)	**735,744**

Source: Golder Associates (2001)

9.2 SITE CHARACTERISATION

The costs for site characterisation for the work carried out by Golder but exclusive of the initial desk study and initial site investigation were £229,500.
This included:

- 33 hand augered boreholes for soil vapour survey
- 14 boreholes used as groundwater monitoring wells
- 5 additional shallow groundwater wells
- 20 rounds of groundwater sampling and laboratory analyses from 20 monitoring wells
- Reports

These costs are assumed to be the same for all remedial options.

9.3 MONITORING

The costs to monitor the site for ten years from 1996 to the end of 2006 are estimated to be £96,000 and included £8,000 which was spent undertaking a tracer test.

9.4 PERMEABLE REACTIVE BARRIER REMEDIATION

The costs for the design and installation of the PRB system were £410,000. This included excavation and disposal of 500 m^3 of contaminated soil, pilot scale evaluation and a completion report.

9.5 COMPARISONS WITH ALTERNATIVE TREATMENTS

The decision to use the PRB was set against cost estimates for alternative treatment technologies as discussed below:

9.5.1 LANDFILLING, AND PUMP AND TREAT USING AIR STRIPPING[1]

Soil removal and disposal costs[2]:	£ 300,000
Capital costs of treatment equipment:	£ 75,000
Operating costs[3]:	£ 360,000
Total Costs:	**£ 735,000**

[1]Assumes 10 year operation
[2]Assumes 2000 m^3 heavily contaminated soil excavated, disposed at a licensed landfill facility and backfilling, at £150/m^3.
[3]Operating costs estimated at £3000/month for pump and treat, air stripping, vapour phase activated carbon treatment, carbon disposal/replacement and monitoring for 10 years.

9.5.2 CONTAINMENT WALL WITH PUMP AND TREAT SYSTEM[4]

Soil removal and disposal[5]	£ 327,500
Containment wall installation[6]	£ 208,000
Capital costs of treatment equipment	£ 50,000
Operating costs[7]	£ 50,000
Total Costs:	**£ 635,500**

[4] Assumes 10 year operation

[5]Assumes 2340 m^3 heavily contaminated soil excavated and disposed at a licensed landfill facility at £140/m^3.

[6]Assume containment wall surrounding area of site 80mx50mx10m

[7]Operation costs estimated at £5,000 per year for pump and treat, liquid phase carbon treatment, carbon disposal, replacement and monitoring for 10 years.

Table 9.2: Summary of costs for different remedial options

	Landfilling and Pump and Treat	**Containment Wall with Pump and Treat**	**Permeable Reactive Barrier**
Site Investigation	£229,500	£229,500	£229,500
Excavation of soil & disposal	£300,000	£327,500	£75,000
Remedial Technology – installation, operation & monitoring	£435,000	£308,000	£431,000
TOTAL	**£964,500**	**£865,000**	**£735,500**

9.6 COST EFFECTIVENESS

The main costs to date associated with this work were capital expenditures, and ongoing monitoring requirements. In assessing cost effectiveness, there are four issues to consider: disposition of the contaminant, installation of the system, ongoing operation of the system, and longevity of the system.

Contaminant disposition can be an important consideration when comparing remedial options, since some options destroy the contaminant, some separate it, while others only contain it. PRB technology is a destructive technology and degrades the contaminant to non-toxic forms. Alternatives such as containment do not destroy the contaminant but rather isolate it from possible receptors leaving the land owner with a contaminant management problem. Landfilling removes the contaminant from the site but transfers the problem of contaminant management to another site and authority. Pump and treat is a long term *ex situ* treatment option which has its own capital and operating costs. Site owners may choose a technology which avoids landfilling or containment as part of a corporate/personal philosophy, in which case PRB technology would provide a viable alternative.

In terms of the installation of the system, Table 9.2 shows that the PRB option was considerably less expensive than either of the landfilling/pump and treat and the containment/pump and treat options. Energy requirements for landfilling/pump and treat and containment/pump and treat will be considerably greater than for PRB because of the increased excavation/ transport and the energy requirements for the pump and treat system.

In terms of ongoing operation, the PRB system has no requirements for man-made energy because the system is passive and *in situ* and is driven by natural conditions. Energy requirements for the pumping of groundwater are derived from renewable sources (wind and solar). Consequently the Fe^0 PRB system is considered to have a very high operational cost effectiveness. Ongoing pump and treat can be operated with renewable energy sources, but will have a non-renewable energy associated with the disposal/replacement of activated carbon.

In terms of system replacement, because PRBs are a relatively new technology, it is not known how long systems will last before plugging occurs, and renovation is required. However the Fe^0 PRB system at Monkstown was designed with a fifty year life on the slurry bentonite wall and reaction vessel, and the Fe^0 is expected to operate for at least 10 -15 years before minor replacement of portions of the iron is necessary. Mechanical breakdown of the supplementary pumping system has not occurred at this time although the system was damaged by weather.

At this time the longevity of the Fe^0 PRB system is considered to be moderate (at least 10-15 years) for minor replacement (iron) and very high (50 years) for major component replacement.

10 CONCLUSIONS

1. The Monkstown site has been operational since 1962 in the manufacture and assembly of electronic components and was purchased by Nortel in the early 1990s. Soil and groundwater contamination consisting predominantly of trichloroethene and its degradation products were discovered during due diligence environmental investigations. Although there was no regulatory requirement to remediate the site at the time, Nortel undertook a voluntary cleanup which consisted of partial excavation and landfilling of contaminated soil and the installation of a zero valent iron (Fe^0) permeable reactive barrier system to treat shallow groundwater in an area of the site known as the eastern car park.

2. The geology at the site consists of more than 18 m of superficial deposits overlying fine to coarse-grained Sherwood Sandstone bedrock of Triassic age. The drift is characterised by a complex succession of stiff, red-brown clayey till, with intercalated, discontinuous lenses of silts, sands, gravels, and peat, overlain by approximately 0.1 to 1.1 m thickness of made ground.

3. Shallow water tables occur at depths ranging between 0.45 and 7.82 mbgl. Shallow, horizontal groundwater flow in the vicinity of the eastern car park is in an easterly to northeasterly direction. Calculated hydraulic conductivities range from 3×10^{-6} metres per second (m/s) in a coarse silty sand to 5×10^{-9} m/s in a clay.

4. Concentrations of TCE in soil during site characterisation ranged from 0.3-1,000 µg/kg. The highest values exceeded the Dutch Target Values but were below Dutch Intervention Levels. Highest concentrations of TCE in groundwater exceeded Dutch Intervention and Target Values and were orders of magnitude higher than other contaminants, with values up to 390,000 µg/L suggesting the presence of free phase TCE.

5. Laboratory scale feasibility studies involving column tests, and using samples of groundwater taken from the site, were used to help design the PRB system. TCE degraded very rapidly with a half life of 1.2 to 3.7 hours, generating c-DCE as an intermediate degradation product with the calculated half life ranging between 12-24 hours. The column test demonstrated that dissolved iron would be expected to occur downgradient from the PRB resulting in the precipitation of siderite and iron oxide.

6. A conceptual model of the site hydrogeology developed by Golder during the site characterisation programme, was modelled using the two dimensional, finite difference, steady-state groundwater flow model FLOWPATH. The purpose of the groundwater flow model was to assist in the design of the PRB system and assess whether the system would operate as designed. The results of the modelling exercise provided an order of magnitude estimation of system parameters and supported the viability of a PRB system at the site. The model predicted that the hydraulic regime at the site would not be significantly altered by the installation of a PRB system and that contaminants would not be diverted around the cut-off wall.

7. Based on field observations, laboratory experiments, modelling and operational constraints, it was decided that a Fe^0 PRB system would be placed at the property boundary of the eastern car park. A cement bentonite cut-off wall would funnel contaminated groundwater to a vertically aligned reaction vessel containing Fe^0.

8. Following the installation of the PRB, a groundwater monitoring programme was established to demonstrate whether the system was operating as designed. The monitoring programme consisted of water level readings and geochemical sampling. Water levels were measured to ensure that the cut-off wall and reactive barrier system had not adversely affected the groundwater conditions. Geochemical sampling of groundwater upgradient, within and downgradient of the reaction vessel was conducted to provide a measure of assessing actual changes in groundwater chemistry of the full scale PRB system against expected results predicted from the laboratory scale column tests, and to demonstrate that discharge from the reactive cell meets design criteria.

9. Monitoring of water levels within the reaction vessel itself indicates periodic reversals of groundwater flow across the reactive cell, making groundwater flow through the reaction vessel difficult to quantify. Groundwater flow through the reaction vessel can be estimated using hydraulic parameters derived from other areas of the site.

10. The major ion chemistry shows the predominant groundwater type upgradient of the PRB to be 'calcium bicarbonate'. Groundwater passing through the reactive cell changes from 'calcium bicarbonate' type to 'magnesium-sodium sulphate-chloride' indicating loss of calcium bicarbonate to calcite precipitation. Contaminant concentrations of TCE have been progressively removed as the groundwater flowed down through the reactive cell.

11. Significant decreases in TCE concentrations in some upgradient wells can be explained by: (i) the removal of highly contaminated material during excavation of the PRB and cut-off wall, although some contaminated material remains; and/or (ii) the tail end of a slug of contamination that moved through the site. The degree to which natural variation, natural attenuation, seasonal fluctuations and disturbance during drilling/excavation affects TCE concentrations in wells cannot be determined from the existing data.

12. Monitoring wells downgradient of the PRB exhibit detectable concentrations of TCE and DCE as a result of residual historic contamination still present in the subsurface. Until such time as levels stabilise and reduce to levels below those found in upgradient wells, downgradient monitoring wells cannot be used to confirm capture of the contaminant plume because their concentrations could mask transgressions of the plume through the cut-off wall.

13. Estimates of groundwater flow through the reactive cell, based on potential capture by the cut-off wall, are estimated at 1-6 m^3/day, with the residence time between 17.4 and 105 hours.

14. Estimates of volumetric flow and residence time along with non-detectable concentrations of TCE and DCE in groundwater leaving the reactor confirm that the reactive cell is operating as designed, meeting the design criterion discharge concentration of 10 µg/L TCE.

15. Investigations by Queens University Belfast, showed that for the Monkstown reactive cell, there was some loss in the activity of the granular iron at the entrance to the reactive cell. Mineralogical observation showed the presence of precipitates in the entrance sample over a very narrow zone only. At a distance of 10-40 cm into the reactive cell centre, the iron activity had improved due to corrosion and increased surface area. The loss of activity in the entry zone is more than offset by the increase in activity in the middle zone.

16. Any significant reduction in reactive cell permeability by precipitation of siderite and calcite leading to restricted flow through the reactive cell, would result in an increase in water levels in upgradient wells. The mineralogical evidence shows that the precipitation reactions occurred over a very narrow zone at the entrance to the reactive cell. Permeability and groundwater flow could be restored by excavating and replacing the "encrusted" iron or by high pressure flushing of the affected area.

17. Based on the field evidence the minimum life expectancy of the iron before any form of maintenance might be required is 10-15 years.

18. No evidence for significant biological fouling within the reactive cell was found.

19. The remediation costs at Monkstown using a PRB system were £735,500. This included site investigation costs, excavation and disposal of 500 m^3 of heavily contaminated soil, capital costs of the system and monitoring projected forward to 10 years. The estimated equivalent costs for alternatives are £964,500 for landfilling/pump and treat and £865,000 for containment/pump and treat.

20. Cost effectiveness of the Fe^0 PRB system was considered in terms of contaminant disposition, installation ongoing operation, and longevity of the system. The PRB system was less expensive to install, and expended less energy than the landfilling/pump and treat and the containment/pump and treat options. In terms of ongoing operation, the system has no requirements for man-made energy and is considered to have a very high operational cost effectiveness. In terms of system replacement, the longevity of the Fe^0 PRB system at Monkstown is expected to be moderate (at least 10-15 years) for minor replacement (iron) and very high (50 years) for any major component replacement.

11 LESSONS LEARNED

The PRB at Monkstown was the first application of Fe^0 PRB technology in Europe and one of the first in the world. Critical analysis of its design and installation provides the opportunity to identify a number of lessons, which were learned from the experience and are discussed below:

1. Involvement of the regulator particularly at an early stage is essential. Although there was no regulatory requirement to carry out work at the site, Nortel, established a positive and open relationship with the regulator, which resulted in confidence by the regulator that the site was being managed in a responsible way. This led the way for open discussions between parties and to the selection of an innovative solution, which was agreed to by both parties.

2. The conditions at Monkstown that were conducive to the application of Fe^0 PRB technology were:

   - chlorinated solvent contamination in shallow groundwater moving offsite
   - low groundwater velocity
   - no evidence of significant biodegradation or other degradation processes for contaminants of concern
   - the presence of a competent aquitard below the contaminated aquifer into which the cut-off wall could be tied
   - the lack of identified discrete sources of contamination

3. The natural head difference across the barrier was less than 0.1 m. Auxiliary pumping was used to re-circulate contaminated groundwater from a local "hot spot" downgradient of the barrier at GA13 to MWU located, immediately upgradient of the barrier. The re-circulation was carried out to take advantage of the unused capacity of the reactive cell. This unused capacity allowed for flexibility in varying the contaminant load and groundwater flow. The re-circulation resulted in an increase in the driving head and a lower groundwater residence time within the PRB. While the residence time was adequate to treat the contaminants at the site, there may be situations where treatment processes will be affected adversely. Re-circulation of contaminated groundwater back into the treatment system requires regulatory approval.

4. The long term chemistry data show a decrease in contaminant concentrations upgradient of the barrier over time. This could be explained by: (i) the excavation of highly contaminated material from the trench which was dug in preparation for the PRB; (ii) the tail end of a slug of contamination that moved through the site; and (iii) other processes such as natural variation, natural attenuation, seasonal fluctuations and disturbances associated with drilling/excavation. While this was not obvious from the initial investigations, it reinforces the need for adequate characterisation and time series sampling.

5. This project illustrates the importance of understanding site specific conditions and the complexities of full-scale natural systems. Reasons for the rapid decline in TCE concentrations in upgradient wells and the gradient reversals observed within the reaction vessel remain unclear. The project also shows the need for proper planning at all stages of site characterisation, remedial planning, installation and monitoring in order to optimise available

funding. It demonstrates the multidisciplinary nature of environmental remediation and the need to involve experienced environmental professionals.

6. The use of PRB at Monkstown is a good example of the cost effective application of a new technology. Site specific conditions led to a novel design for the reactive cell. This project illustrates the importance of adequate site characterisation, laboratory studies, flexibility of approach and ongoing monitoring in the design and implementation of remedial systems.

12 GLOSSARY OF TERMS

acetone
a volatile fragrant flammable liquid ketone C_3H_6O used chiefly as a solvent and in inorganic synthesis.

anthropogenic
of, relating to, or resulting from the influence of human beings on nature.

aquiclude
a subsurface unit of low permeability which restricts the flow of groundwater.

aquifer
a subsurface permeable unit which is capable of transmitting significant quantities of groundwater.

bioaugmentation
the addition of naturally occurring microbes.

biofouling
the gradual accumulation of waterborne organisms (as bacteria and protozoa) on the surfaces of engineering structures in water that contributes to corrosion of the structures and to a decrease in the efficiency of moving parts.

butanone
a colourless flammable water-soluble liquid, $CH_3COC_2H_5$, commonly used as a solvent.

cable percussion rigs
suited for drilling boreholes in areas contaminated by hazardous substances, because they do not use any circulation fluids that could spread contamination. The rigs operate by repeatedly lifting and dropping a heavy string of drilling tools attached to a cable into the borehole.

carbon isotope
an isotope can be defined as one of two or more forms of an element that differ in relative atomic mass and nuclear properties, but are chemically identical. The isotopes of carbon have the same number of protons in their nuclei but different numbers of neutrons. Natural carbon is composed of three isotopes: ^{12}C making up about 98.9%; ^{13}C about 1.1%; and ^{14}C whose amount is negligible, but which is detectable because it is radioactive. The relative abundance of these isotopes varies and the study of this variation is an important tool in geologic research as isotopic fractionation refers to the fluctuation in the carbon isotope ratios as a result of natural biochemical processes as a function of their atomic mass.

cement bentonite
mixtures of Portland cement with 2 to 10 % bentonite clay are the recommended sealant material to use when decommissioning a contaminated well because, unlike neat cement that shrinks and can crack upon curing, cement-bentonite grout swells and remains plastic when cured which in turn creates a superior seal.

clean up criteria
numerical concentration values assigned to different contaminants which need to be met in order for a site to be considered remediated. These criteria often form part of regulatory legislation.

column treatability tests
laboratory based studies which involve passing a liquid through a column packed with porous medium. In the case of Monkstown, a column treatability study was conducted on granular iron using groundwater flow rates and chemistry expected to occur at the site to establish design specifications for the PRB system.

contamination
the consequence of any substance introduced into air, water or to the ground which has the effect of rendering them toxic or otherwise harmful.

Cr(III)
the trivalent chromium ion does form compounds that are poisonous, but due to their low solubility create little risk.

Cr(VI)
the hexavalent chromium is a human carcinogen that exists in soils and natural waters predominantly as a soluble anion that may form via oxidation of soluble and insoluble forms of nontoxic Cr(III) in soils amended with industrial waste materials. The hexavalent chromium ion is acidic and forms soluble chromates and dichromates. The main effects of chromates are observed on the skin and mucous membranes and they are also very toxic to aquatic plant and animal life.

denitrification
the loss or removal of nitrogen or nitrogen compounds. More specifically, the reduction of nitrates or nitrites commonly by bacteria (as in soil) that usually results in the formation of nitrogen gas.

downgradient
describes the zone which is located relative to and away from a fixed point in the direction of groundwater flow.

falling head slug tests
used in the hydraulic characterisation of a site. A slug test involves the insertion or removal of a known volume of water or the displacement of water by a solid object. A falling head slug test involves introducing an object or volume of water (the slug) into the well and recording how long it takes the water to return to its initial level. Water levels versus time are recorded during this "falling head" portion of the test.

filter piles
the columnar sections filled with sand, which were installed on either side of the PRB reaction vessel and are hydraulically linked to both the vessel and the natural/man-made ground.

hardstanding
an area with a hard impermeable surface such as concrete.

hydraulic conductivity (K)
the measure of how easily a medium can transmit a specified fluid. In groundwater terms it relates to an aquifer's ability to transmit water and is often expressed in terms of metres/sec.

in situ remediation
treatment of contamination in place, without removal.

inorganic
of, relating to, or dealt with by a branch of chemistry concerned with substances not usually classed as organic. Apart from such analytes as carbonates and cyanides, inorganic chemicals are those that contain no carbon.

intercalated
inserted between or among existing elements or layers.

organic
chemically, a substance containing carbon in the molecule (with the exception of carbonates and cyanide).

penetrometer
an instrument for measuring firmness or consistency (as of soil).

permeable reactive barrier (PRB)
a PRB is an in situ passive treatment system used to remediate contaminated fluids such as groundwater. It consists of a permeable wall of reactive material which is installed across the flow path of the contaminated fluid. As the fluid flows through the permeable barrier, the contaminant comes into contact with the reactive material and depending on the nature of the reactive material, is degraded to non or less toxic forms or its rate of transport is retarded.

potentiometric surface
a hypothetical surface defined by the level to which water in a confined aquifer rises in observation boreholes.

reaction cell
the lower portion of the reaction vessel containing the zero valent iron.

reaction vessel
the tubular steel container which houses the zero valent iron and monitoring wells R5 to R1, which was inserted into the cement slurry cut-off wall, and which along with the cut-off wall forms the PRB system at Monkstown.

reductive dechlorination
specific chemical process for the removal of chlorine from contaminant compounds (see reductive dehalogenation below).

reductive dehalogenation
a series of chemical reactions in which a halogen (fluorine, chlorine, bromine, iodine or astatine) is removed.

remediation
the process of making a site fit-for-purpose through destruction, removal or containment of contaminants.

retardation
a process of slowing down the movement of contaminants through natural or engineered processes.

sorption
processes including adsorption and absorption, by which contaminants attach themselves to solid particles, thereby retarding their transport or movement.

till
unstratified glacial drift which can consist of mixed clay, silt, sand, gravel, and boulders.

unconfined aquifer
a near surface aquifer at atmospheric pressure in which the top of the aquifer is defined by the water table.

upgradient
describes the zone which is located away from a fixed point in the opposite direction of groundwater flow.

valence state
atoms are assigned numbers, called valence numbers or oxidation numbers, which range in value from -4 through 0 to +7. These numbers describe the valence or oxidation state and relate to the combining behaviour of the atoms in chemical reactions, particularly oxidation-reduction reactions.

zero valent iron
the elemental form of iron, Fe^0, with a valence state of zero.

REFERENCES

Bill, M., C. Schüth, J.A.C. Barth and R.M. Kalin. 2000. Carbon Isotope Transfer during Abiotic Reductive Dechlorination of Trichloroethene (TCE). Chemosphere 44, 1281-1286.

Council of the European Community. 1980. Council Directive of 17 December 1979 on the Protection of Groundwater Against Pollution Caused by Certain Dangerous Substances (80/68/EEC), Official Journal of the European Communities, L20, 43-47.

Freeze, R. A., J.A. Cherry. 1979. Groundwater. Prentice Hall.

Gavaskar, A.R., N. Gupta, B.M. Sass, R.J. Janosy and D. O'Sullivan. 1998. Permeable Barriers for Groundwater Remediation - Design, Construction, and Monitoring. Battelle Press.

Gillham, R.W., and S.F. O'Hannesin. 1992. Metal-Catalyzed Abiotic Degradation of Halogenated Organic Compounds. IAH Conference: Modern Trends in Hydrogeology. Hamilton, Ontario, May 10-13, pp.94-103.

Gillham, R.W., S.F. O'Hannesin and W.S. Orth. 1993. Metal Enhanced Abiotic Degradation of Halogenated Aliphatics: Laboratory Tests and Field Trials. Proceedings of the 6th Annual Environmental Management and Technical Conference/HazMat Central Conference, Rosemont, Illinois, pp.440-461.

Gillham, R.W. 1993. Cleaning Halogenated Contaminants from Groundwater. U.S. Patent No. 5, 266, 213, Nov.30.

Gillham, R.W., and S.F. O'Hannesin. 1994. Enhanced Degradation of Halogenated Aliphatics by Zero-Valent Iron. Ground Water 32, 958-967.

Gillham, R.W. 2001. International Containment and Remediation Technologies Conference and Exhibition, Orlando, Florida, 11-13 June 2001.

Golder Associates. 1994. Numerical Modelling of the Proposed Remediation Scheme at Monkstown.

Golder Associates. 1994. Northern Telecom Environmental Investigations at Monkstown.

Golder Associates. 1996. Completion Report for the Remediation Works at Nortel Ltd, Monkstown, Northern Ireland.

Golder Associates. 2001. Monitoring Results.

Health and Safety Commission. 1994. Managing Construction for Health and Safety Construction (Design and Management) Regulations.

Hvorslev, J. 1951. Time Lag and Soil Permeability in Groundwater Observations. US Waterways Experiment Station Bulletin 36. Vicksburg.

ICE, CIRIA, BRE and DETR. 1999. Specification for the Construction of Slurry Trench Cut-off walls or Barriers to Pollution Migration. Thames Telford.

Institute for Groundwater Research. 1995. Report on the Column Treatability Tests Conducted on Groundwater from Northern Telecom, Monkstown Facility.

Jefferis, S.A., G.H. Norris and A.O. Thomas. 1997. Contaminant Barriers: from Passive Contaminant to Reactive Treatment Zones, Fourteenth International Conference on Soil Mechanics and Geotechnical Engineering, Hamburg, September.

Ministry for Housing, Spatial Planning and the Environment. 1994. Dutch Soil and Groundwater Values.

Ministry for Housing, Spatial Planning and the Environment. 2000. Dutch Soil and Groundwater Values.

NATO/CCMS. 1998. Pilot Study - Evaluation of Demonstrated and Emerging Technologies for the Treatment of Contaminated Land and Groundwater (Phase III) - Special Session "Treatment Walls and Permeable Reactive Barriers" Number 229, EPA 542-R-98-003.

O'Hannesin, S. 1999. Written correspondence from Envirometal Technologies Inc. to CL:AIRE - "Follow up information". Nov.

Powell & Associates Science Services - Website www.powellassociates.com.

Reynolds, G.W., J.T. Hoff, and R.W. Gillham. 1990. Sampling Bias Caused by Materials Used to Monitor Halocarbons in Groundwater. Environ. Sci.Technol., 24(1), 135-142.

Robins, N.S. 1994. Hydrogeological Map of Northern Ireland, 1:250,000, British Geological Survey and Environment Service, Department of the Environment for Northern Ireland.

Schüth, C. M., Bill, J. Barth, G. Slater, R.M. Kalin. 2000. Long term performance of iron reactive barriers - rates and carbon isotope fractionation in batch-experiments with original samples. Paper submitted for publication.

Senzaki, T. and Y. Kumagai. 1988a. Removal of Chlorinated Organic Compounds from Wastewater by Reduction Process: Treatment of 1,1,2,2-Tetrachloroethane with Iron Powder. Kogyo Yosui 357, 2-7 (in Japanese).

Senzaki, T. and Y. Kumagai. 1988b. Removal of Chlorinated Organic Compounds from Wastewater by Reduction Process: II. Treatment of Tetrachloroethane with Iron Powder. Kogyo Yosui 369, 19-25 (in Japanese).

Senzaki, T. 1988. Removal of Chlorinated Organic Compounds from Wastewater by Reduction Process: III. Treatment of Tetrachloroethane with Iron Powder II. Kogyo Yosui 391: 29-35 (in Japanese).

Standard Methods for the Examination of Water and Wastewater. 1989. 17th Edition. American Public Health Association, Washington, DC.

Sweeny, K.H., and J.R. Fischer. 1972. Reductive Degradation of Halogenated Pesticides. U.S.Patent No. 3,640,821.

Sweeny, K.H. 1981a. The Reductive Treatment of Industrial Wastewaters: I. Process Description. In G.F. Bennett (Ed.), American Institute of Chemical Engineers, Symposium Series, Water-1980, 77 (209): 67-71.

Sweeny, K.H. 1981b. The Reductive Treatment of Industrial Wastewaters: II.Process Description. In G.F. Bennett (Ed.), American Institute of Chemical Engineers, Symposium Series, Water-1980, 77 (209): 72-88.

U.S. Environmental Protection Agency. 1982. Methods for organic chemical analysis of municipal and industrial wastewater; EPA-600/4-82-057, J.E. Longbottom and J.J. Lichtenberg (eds.); Cincinnati, Ohio; Appendix A.

U.S. Environmental Protection Agency. 1995. In Situ Remediation Technology Status Report: Treatment Walls. EPA 542-K-94-004.U.S. EPA, Office of Solid Waste and Emergency Response, April.

U.S. Environmental Protection Agency. 1998. Permeable Reactive Barrier Technologies for Contaminant Remediation. EPA 600-R-98-125. U.S. EPA. Remedial Technology Development Forum, September.

U.S. Environmental Protection Agency. 1999. An In Situ Permeable Reactive Barrier for the Treatment of Hexavalent Chromium and Trichloroethylene in Groundwater: Volume 1 - Design and Installation. EPA 600-R-99-095a.U.S. EPA.Office of Research and Development, September.

U.S. Environmental Protection Agency. 1999. An In Situ Permeable Reactive Barrier for the Treatment of Hexavalent Chromium and Trichloroethylene in Groundwater: Volume 2 - Performance Monitoring. EPA 600-R-99-095b.U.S. EPA.Office of Research and Development, September.

U.S. Environmental Protection Agency. 1999. An In Situ Permeable Reactive Barrier for the Treatment of Hexavalent Chromium and Trichloroethylene in Groundwater: Volume 3 - Multicomponent Reactive Transport Modeling. EPA 600-R-99-095c.U.S. EPA. Office of Research and Development, September.

WESA. 1991. Preliminary Environmental Review Northern Telecom Ltd, Northern Ireland (Monkstown).

WESA. 1993. Hydrogeological Investigation of Northern Telecom Ltd, Northern Ireland (Monkstown)

APPENDICES

APPENDIX 1 - REPRESENTATIVE BOREHOLE LOGS

BH1
BH2
BH6
BH7
BH8
BH10
BH11
BH12
BH16
BH17
BH18
BH19
BH20
BH21

GA3
GA4
GA5
GA6
GA7
GA8
GA9
GA10
GA11
GA12
GA13
GA14
GA16
GA17
GA18
GA19
GA20
GA21

FIGURE:	RECORD OF TEST HOLE	DESIGNATION BH 1	COMPLETION DATE March 18, 1992

PROJECT: Monkstown N.T.E.
PROJECT NO.: B1007
DRILLING METHODS: Air Rotary Drill
SUPERVISOR: D. Corbett
DRILLING CONTRACTOR: Stratex Ltd.

DEPTH METRES	ELEVATION METRES	STRATIGRAPHY	LOG	INSTRUMENTATION	TYPE	INTERVAL	HNu (ppmv)
				cap & steel cover			
0		grassed surface					
		topsoil; dark brown & silty		concrete	SS0.4 -0.9		5.0
		gravel with silt matrix		0.2			
2		red-brown oxidized silt till with a trace of gravel and coarse sand; gravel is well rounded			SS3.0 -3.5		3.5
4				bentonite/cement grout			
6		becomes stiffer, higher clay content			SS6.0 -6.3		
8				50mm I.D. HDPE pipe			
10		cobble sized clasts; becomes drier with depth			SS9.0 -9.5		3.5
12					SS12.0 -12.5		
14		gravel and cobble clasts become more numerous - no obvious fracturing but still oxidized		14.0			
16					SS15.0 -15.5		
18				Natural Backfill	SS18.0 -18.5		5.0
20				20.5 Bentonite			
22				21.5 3mm silica sand			
24		24.5 oxidized mudstone		24.5			
		24.5m - END OF HOLE		200 mm hole 1.0m x 50mm HDPE screen wrapped in geotextile			
26							
28							

WATER AND EARTH SCIENCE ASSOCIATES LTD.
116 Cheyne Walk London, England. U.K.

FIGURE:

RECORD OF TEST HOLE

DESIGNATION: BH 2

COMPLETION DATE: March 19, 1992

PROJECT: Monkstown N.T.E.

PROJECT NO.: B1007

DRILLING METHODS: Air Rotary Drill

SUPERVISOR: D. Corbett

DRILLING CONTRACTOR: Stratex Ltd.

DEPTH METRES	ELEVATION METRES	STRATIGRAPHY	LOG	INSTRUMENTATION	TYPE	INTERVAL	HNu (ppmv)
0		Asphalt		cap & steel cover			
0.5		Fill material 0.5		concrete 0.3			
1.0		Silty till with some clay and a trace of gravel			SS0.5 −1.0		
1.5				bentonite/cement grout			
2.0				50mm I.D. HDPE pipe			
2.5							
3.0				3.0	SS3.0 −3.5		
3.5		Becomes stiffer with higher proportion of clay becomes drier and brown at depth		bentonite			
4.0				4.0			
4.5				3mm silica sand			
5.0		gravel clasts larger					
5.5				1.0m x 50mm HDPE screen wrapped in geotextile			
6.0		6.0		6.0	SS6.0 −6.5		
		6.0m – END OF HOLE		200 mm hole			
6.5							
7.0							

WATER AND EARTH SCIENCE ASSOCIATES LTD.
116 Cheyne Walk London, England, U.K.

FIGURE:	RECORD OF TEST HOLE	DESIGNATION	COMPLETION DATE
		BH 6	March 20, 1992

PROJECT:	Monkstown N.T.E.	DRILLING METHODS:	Air Rotary Drill
PROJECT NO.:	B1007	SUPERVISOR:	A. Huntley
		DRILLING CONTRACTOR:	Stratex Ltd.

DEPTH METRES	ELEVATION METRES	STRATIGRAPHY	LOG	INSTRUMENTATION	TYPE	INTERVAL	HNu (ppmv)
				cap & steel cover			
0		Concrete surface					
		Fill material		concrete			
0.5		0.5		0.3			
1.0		stiff brown clayey silt till; occasional seams of fine-grained sand		bentonite/cement grout			
1.5					SS0.9 −1.2		
2.0				50mm I.D. HDPE pipe			
2.5							
3.0				3.2	SS3.0 −3.5		
3.5				bentonite			
4.0				4.2			
4.5				3mm silica sand			
5.0							
5.5		5.5		1.0m x 50mm HDPE screen wrapped in geotextile			
6.0		coarse-grained sand with some silt					
6.5		6.6		6.6			
		6.6m − END OF HOLE		200 mm hole	SS6.6 −7.0		
7.0							

WATER AND EARTH SCIENCE ASSOCIATES LTD.
116 Cheyne Walk London, England, U.K.

FIGURE:	RECORD OF TEST HOLE	DESIGNATION BH 7	COMPLETION DATE March 23, 1992
PROJECT: Monkstown N.T.E.		DRILLING METHODS: Air Rotary Drill	
PROJECT NO.: B1007		SUPERVISOR: D. Corbett	
		DRILLING CONTRACTOR: Stratex Ltd.	

DEPTH METRES	ELEVATION METRES	STRATIGRAPHY	LOG	INSTRUMENTATION	TYPE	INTERVAL	HNu (ppmv)
0		Asphalt surface		cap & steel cover			
		Fill material		concrete			
0.5		0.4		0.3			
1.0		stiff brown clayey silt till with a trace of gravel; saturated			SS0.4 -0.9		11.8
1.5				bentonite/cement grout			
2.0				50mm I.D. HDPE pipe			
2.5							
3.0					SS3.0 -3.5		9.5
3.5							
4.0							
4.5				4.5			
5.0				bentonite			
5.5				5.5			
6.0		clay becoming more stiff					
6.5				3mm silica sand	SS6.0 -6.5		7.8
7.0							
7.5		7.5		7.5			
		7.5m - END OF HOLE		1.0m x 50mm HDPE screen wrapped in geotextile; 200 mm hole	SS7.5 -8.0		6.5
8.0							

WATER AND EARTH SCIENCE ASSOCIATES LTD.
116 Cheyne Walk London, England, U.K.

FIGURE:	RECORD OF TEST HOLE	DESIGNATION BH 8	COMPLETION DATE March 23, 1992
PROJECT: Monkstown N.T.E.		DRILLING METHODS: Air Rotary Drill	
PROJECT NO.: B1007		SUPERVISOR: D. Corbett	
		DRILLING CONTRACTOR: Stratex Ltd.	

DEPTH METRES	ELEVATION METRES	STRATIGRAPHY	LOG	INSTRUMENTATION	TYPE	INTERVAL	HNu (ppmv)
0		Asphalt		cap & steel cover			
0.5		Fill material		concrete; 0.3			
1.0		1.0		bentonite/cement grout			
1.5		silty clay till with a trace of gravel; saturated			SS1.0 −1.5		6.5
2.0		grey/green oxidation and mottling		50mm I.D. HDPE pipe			
2.5							
3.0		clay becomes drier and more stiff		3.0 bentonite	SS3.0 −3.5		5.0
3.5							
4.0				4.0			
4.5				3mm silica sand			
5.0							
5.5				1.0m x 50mm HDPE screen wrapped in geotextile			
6.0		6.0 6.0m − END OF HOLE		6.0 200 mm hole	SS6.0 −6.5		7.0
6.5							
7.0							

WATER AND EARTH SCIENCE ASSOCIATES LTD.
118 Cheyne Walk London, England, U.K.

FIGURE:	RECORD OF TEST HOLE	DESIGNATION BH 10	COMPLETION DATE March 24, 1992

PROJECT: Monkstown N.T.E.

PROJECT NO.: B1007

DRILLING METHODS: Air Rotary Drill

SUPERVISOR: D. Corbett

DRILLING CONTRACTOR: Stratex Ltd.

DEPTH METRES	ELEVATION METRES	STRATIGRAPHY	LOG	INSTRUMENTATION	TYPE	INTERVAL	HNu (ppmv)
				cap & steel cover			
0		Concrete surface					
		Fill material		concrete 0.2			
0.5							
1.0		1.0		bentonite/cement grout			
		brown stiff silty clay till with a trace of fine gravel; moist			SS1.0 -1.5		
1.5							
2.0				50mm I.D. HDPE pipe			
2.5							
3.0				3.0 bentonite	SS3.0 -3.5		5.5
3.5							
4.0		becomes more stiff and dry with depth; more gravel		4.0			
4.5				3mm silica sand			
5.0							
5.5				1.0m x 50mm HDPE screen wrapped in geotextile			
6.0		6.0		6.0	SS6.0 -6.5		3.5
		6.0m - END OF HOLE		200 mm hole			
6.5							
7.0							

WATER AND EARTH SCIENCE ASSOCIATES LTD.
116 Cheyne Walk London, England, U.K.

FIGURE:	RECORD OF TEST HOLE	DESIGNATION BH 11	COMPLETION DATE March 24, 1992

PROJECT: Monkstown N.T.E.

PROJECT NO.: 81007

DRILLING METHODS: Air Rotary Drill

SUPERVISOR: D. Corbett

DRILLING CONTRACTOR: Stratex Ltd.

DEPTH METRES	ELEVATION METRES	STRATIGRAPHY	LOG	INSTRUMENTATION	TYPE	INTERVAL	HNu (ppmv)
				cap & steel cover			
0		Asphalt surface					
		Fill material		concrete 0.2			
0.5							
1.0		1.0		bentonite/cement grout			
		brown stiff silty clay till with a trace of gravel			SS1.1 −1.6		5.0
1.5							
2.0				50mm I.D. HDPE pipe			
2.5							
3.0				3.0 bentonite	SS3.0 −3.5		6.5
3.5							
4.0		Becoming stiffer with depth		4.0			
4.5				3mm silica sand			
5.0							
5.5				1.0m x 50mm HDPE screen wrapped in geotextile			
6.0		6.0		6.0	SS6.0 −6.5		
		6.0m – END OF HOLE		200 mm hole			
6.5							
7.0							

WATER AND EARTH SCIENCE ASSOCIATES LTD.
115 Cheyne Walk London, England, U.K.

FIGURE:	RECORD OF TEST HOLE	DESIGNATION BH 12	COMPLETION DATE March 25, 1992

PROJECT:	Monkstown N.T.E.	DRILLING METHODS:	Air Rotary Drill
PROJECT NO.:	B1007	SUPERVISOR:	D. Corbett
		DRILLING CONTRACTOR:	Stratex Ltd.

DEPTH METRES	ELEVATION METRES	STRATIGRAPHY	LOG	INSTRUMENTATION	TYPE	INTERVAL	HNu (ppmv)
0		Grassed surface		cap & steel cover			
0.5		0.5 topsoil: dark brown & silty		concrete 0.2			
1.0		brown silty clay till with some sand; moist			SS0.5 −1.0		
1.5				bentonite/cement grout			
2.0				1.7 bentonite			
2.5							
3.0				2.7	SS3.0 −3.5		4.8
3.5		3.4 coarse sand and gravel; saturated		50mm I.D. HDPE pipe			
4.0				3mm silica sand			
4.5				4.7			
5.0		5.0 silty clay		5.0			
5.5		5.0m – END OF HOLE		200 mm hole; 1.0m x 50mm HDPE screen wrapped in geotextile	SS5.0 −5.5		
6.0							
6.5							
7.0							

WATER AND EARTH SCIENCE ASSOCIATES LTD.
116 Cheyne Walk London, England, U.K.

FIGURE:	RECORD OF TEST HOLE	DESIGNATION BH 16	COMPLETION DATE May 26, 1992

PROJECT: Monkstown N.T.E.
PROJECT NO.: B1007
DRILLING METHODS: Air Rotary Drill
SUPERVISOR: D. Corbett
DRILLING CONTRACTOR: Stratex Ltd.

DEPTH METRES	ELEVATION METRES	STRATIGRAPHY	LOG	INSTRUMENTATION	TYPE	INTERVAL	N VALUE (ppmv)
				50mm HDPE Pipe			
0		Grass					
		Topsoil: dark brown & silty		concrete & steel cover			
0.5		0.5		0.3			
		Made ground: dark brown; sandy			SS0.5 -1.0		8.6
1.0				bentonite/cement grout			
		1.3					
1.5		Silty till with some clay & a trace of gravel; silt matrix with occasional pebbles; oxidized red-brown			SS1.3 -1.8		1.4
2.0		Peat layer from 1.65–2.80m; black, organic, unconsolidated					
2.5							
3.0				3.0 bentonite	U4 6.0 -6.2		1.8
3.5							
4.0		Till becoming less weathered & more stiff; more pebbles & cobbles; red-brown		4.0			
4.5				3mm silica sand			
5.0				5.0			
5.5				1.0m x 50mm dia. HDPE screen wrapped in geotextile			
6.0		6.0 Water at 5.8m		6.0	U46.0 -6.2		3.6
		6.0m – END OF HOLE		200 mm hole			
6.5							
7.0							

WATER AND EARTH SCIENCE ASSOCIATES LTD.
116 Cheyne Walk London, England, U.K.

FIGURE:	RECORD OF TEST HOLE	DESIGNATION BH 17	COMPLETION DATE May 26, 1992

PROJECT: Monkstown N.T.E.

PROJECT NO.: B1007

DRILLING METHODS: Air Rotary Drill

SUPERVISOR: D. Corbett

DRILLING CONTRACTOR: Stratex Ltd.

DEPTH METRES	ELEVATION METRES	STRATIGRAPHY	LOG	INSTRUMENTATION	TYPE	INTERVAL	N VALUE (ppmv)
0		Grass		50mm HDPE Pipe			
0.5		Topsoil & made ground; dark brown & sandy		concrete & steel cover; 0.3			
1.0		0.8			U4 0.8 −1.2		3.8
1.5		Silty till with some clay & a trace of gravel; silt matrix with occasional pebbles; oxidized dark brown & red		bentonite/cement grout			
2.0							
2.5							
3.0				3.0 bentonite	U4 3.0 −3.2		4.6
3.5							
4.0		Till becoming less weathered & stiffer with depth; numerous pebbles and cobbles, red-brown		4.0 3mm silica sand			
4.5							
5.0							
5.5				1.0m x 50mm dia. HDPE screen wrapped in geotextile			
6.0		6.0		6.0	U4 6.0 −6.2		4.2
		6.0m – END OF HOLE		200 mm hole			
6.5							
7.0							

WATER AND EARTH SCIENCE ASSOCIATES LTD.
116 Cheyne Walk London, England, U.K.

FIGURE:	RECORD OF TEST HOLE	DESIGNATION BH 18	COMPLETION DATE May 26, 1992

PROJECT: Monkstown N.T.E.
PROJECT NO.: B1007

DRILLING METHODS: Air Rotary Drill
SUPERVISOR: D. Corbett
DRILLING CONTRACTOR: Stratex Ltd.

DEPTH METRES	ELEVATION METRES	STRATIGRAPHY	LOG	INSTRUMENTATION	TYPE	INTERVAL	N VALUE (ppmv)
				50mm HDPE Pipe			
0		Grass		concrete & steel cover			
0.5		Topsoil & made ground; dark brown & sandy		0.3			
1.0		0.8 Silty till with some clay & a trace of gravel; silt matrix with occasional pebbles; oxidized dark brown & red		bentonite/cement grout	U4 0.8 –1.0		116
1.5							
2.0							
2.5							
3.0				3.0 bentonite	U4 3.0 –3.2		17.4
3.5							
4.0		Till becoming less weathered & stiffer with depth; numerous pebbles and cobbles, red–brown		4.0 3mm silica sand			
4.5							
5.0				1.0m x 50mm dia. HDPE screen wrapped in geotextile			
5.5							
6.0		6.0 — 6.0m – END OF HOLE		6.0; 200 mm hole	U4 6.0 –6.2		4.4
6.5							
7.0							

WATER AND EARTH SCIENCE ASSOCIATES LTD.
116 Cheyne Walk London, England, U.K.

FIGURE:	RECORD OF TEST HOLE	DESIGNATION BH 19	COMPLETION DATE May 27, 1992

PROJECT:	Monkstown N.T.E.	DRILLING METHODS:	Air Rotary Drill
PROJECT NO.:	B1007	SUPERVISOR:	D. Corbett
		DRILLING CONTRACTOR:	Stratex Ltd.

DEPTH METRES	ELEVATION METRES	STRATIGRAPHY	LOG	INSTRUMENTATION	TYPE	INTERVAL	N VALUE (ppmv)
0		Asphalt		50mm HDPE Pipe			
0.5		Asphalt & fill material: dark brown; sandy (0.8)		concrete & steel cover; 0.3	U4 0.5 –0.8		44
1.0 – 2.5		Silty till with some clay & a trace of gravel; silt matrix with occasional pebbles; oxidized dark brown & red		bentonite/cement grout			
3.0				3.0 bentonite	U4 3.0 –3.2		36
4.0		Till becoming less weathered & stiffer with depth; numerous pebbles and cobbles, red-brown		4.0			
4.5 – 5.0				3mm silica sand			
5.5		Fine-grained sand & gravel lense at 5.7m; Water at 5.7m		1.0m x 50mm dia. HDPE screen wrapped in geotextile			
6.0		6.0		6.0	U4 6.0 –6.2		230
		6.0m – END OF HOLE		200 mm hole			
6.5							
7.0							

WATER AND EARTH SCIENCE ASSOCIATES LTD.
116 Cheyne Walk London, England, U.K.

FIGURE:	RECORD OF TEST HOLE	DESIGNATION BH 20	COMPLETION DATE May 27, 1992

PROJECT: Monkstown N.T.E.
PROJECT NO.: B1007

DRILLING METHODS: Air Rotary Drill
SUPERVISOR: D. Corbett
DRILLING CONTRACTOR: Stratex Ltd.

DEPTH METRES	ELEVATION METRES	STRATIGRAPHY	INSTRUMENTATION	TYPE	INTERVAL	N VALUE (ppmv)
0		Asphalt	50mm HDPE Pipe			
0–0.8		Asphalt & fill material; dark brown & sandy	concrete & steel cover (0–0.3)			
0.8–6.0		Silty till with some clay & a trace of gravel; silt matrix with occasional pebbles; oxidized dark brown & red	0.3–3.0 bentonite/cement grout	U4 0.8 −1.0	0.8–1.0	0.8
				U4 1.7 −1.9	1.7–1.9	1.4
			3.0–4.0 bentonite	U4 3.0 −3.2	3.0–3.2	4.6
		Till becoming less weathered & stiffer with depth;numerous pebbles and cobbles, red-brown	4.0–6.0 3mm silica sand; 1.0m x 50mm dia. HDPE screen wrapped in geotextile			
6.0		6.0m – END OF HOLE	6.0; 200 mm hole	U4 6.0 −6.2	6.0–6.2	3.8

WATER AND EARTH SCIENCE ASSOCIATES LTD.
116 Cheyne Walk London, England, U.K.

FIGURE:	RECORD OF TEST HOLE	DESIGNATION BH 21	COMPLETION DATE May 27, 1992

PROJECT:	Monkstown N.T.E.
PROJECT NO.:	B1007
DRILLING METHODS:	Air Rotary Drill
SUPERVISOR:	D. Corbett
DRILLING CONTRACTOR:	Stratex Ltd.

DEPTH METRES	ELEVATION METRES	STRATIGRAPHY	LOG	INSTRUMENTATION	TYPE	INTERVAL	N VALUE (ppmv)
0		Asphalt		50mm HDPE Pipe			
0–0.8		Asphalt & fill material; dark brown & sandy		concrete & steel cover (to 0.3)	U4 0.8 –1.0		0.6
0.8–5.2		Silty till with some clay & a trace of gravel; silt matrix with occasional pebbles; oxidized dark brown & red		bentonite/cement grout (0.3–3.0)	U4 3.0 –3.2		1.1
		Till becoming less weathered & stiffer with depth; numerous pebbles and cobbles, red-brown		bentonite (3.0–4.0); 3mm silica sand (4.0–6.0)			
5.2–6.0		Water at 5.2m. Coarse sand & fine gravel: stratified; some silt		1.0m x 50mm dia. HDPE screen wrapped in geotextile	U4 5.9 –6.1		1.2
6.0		6.0m – END OF HOLE		6.0; 200 mm hole			

Depth scale: 0, 0.5, 1.0, 1.5, 2.0, 2.5, 3.0, 3.5, 4.0, 4.5, 5.0, 5.5, 6.0, 6.5, 7.0

WATER AND EARTH SCIENCE ASSOCIATES LTD.
115 Cheyne Walk London, England, U.K.

Project Monkstown		BOREHOLE No. GA3
Client Northern Telecom	Contractor Stratex	Sheet 1 of 1
Rig Type Shell & Auger	Location (See Borehole Location Map)	Driller CW Engineer KAF
Hole Diameter 6 1/2 inch	Ground Surface Elevation 6 1/2 inch mOD	Date Started 10/03/94
Completion Depth 7.0m bgl	Casing Details Temporary casing during drilling	Date Completed 10/03/94

Downhole Depth	Borehole Progress	Depth to Water	Samples/Tests Depth/m From	To	No.	Type	Symbolic Log	Elevation, m OD	Depth, m (thickness m)	Geological Description	Installation Details
									0.10	Dark brown organic rich SOIL with grass roots etc.	
									0.76	MADE GROUND - soft to firm, dark brown clay with organic remians (twigs, roots, etc) and fragments of brick and glass	
1			1.0	1.2		D			1.00	Firm yellow-brown CLAY	
									1.40	Firm dark grey/black CLAY with abundant organic fragments	
									1.80	Firm pale grey CLAY	
2			1.8	2.0		D			2.90	Soft to firm, dark grey peaty SILT with abundant organic fragments and small white gastraped shells	
3			2.8	3.0		D			3.30	Soft, wet grey CLAY with black organic rich patches	
4			3.8	4.0		D				Stiff, dry red-brown CLAY till containing subangular clasts of mixed size and lithology. Sizes range from sand to cobble size. Lithologies include flints and dolerite/basalt	
5			4.8	5.0		D				Till becomes sandier at 5m	
			5.3	5.5		D			5.7		
			5.5	5.6		D					
6										Loose, wet, coarse SAND and gravel with small amounts of silt and clay	
7			6.8	7.0		D			7.00		
8											
9											
10											

Remarks Grass soil and some clays later excavated to lay foundation for new carpark
D = Disturbed sample for FID screening
3m screen w. gravel pack + 4m plain casing + (4-7m bgl)
1m bentonite plug (3-4m bgl)
2m grout (1-3m bgl)
1m concrete and headworks (0-1m bgl)

Scale 1:50	Golder Associates	Cable Percussion/Rotary Openhole Record	Project No. 93525199.4000

Project Monkstown		BOREHOLE No. GA4
Client Northern Telecom	Contractor Stratex	Sheet 1 of 1
Rig Type Shell & Auger	Location (See Borehole Location Map)	Driller MG Engineer KF
Hole Diameter 6 1/2 inch	Ground Surface Elevation 40.4 mOD	Date Started 10/03/94
Completion Depth 7.0m bgl	Casing Details Temporary casing during drilling	Date Completed 10/03/94

Downhole Depth	Borehole Progress	Depth to Water	Samples/Tests Depth/m From	To	No.	Type	Symbolic Log	Elevation, m OD	Depth, m (thickness m)	Geological Description	Installation Details
									0.10	Dark brown SOIL with roots, grasses etc.	
										Firm to stiff, red-brown CLAY till with sub angular clasts of sands, gravels pebble and, occasionally, cobble size	
			0.8	1.0		D					
1											
			1.8	2.0		D					
2									(2.20)	(Large cobble (chiselling)	
			2.8	3.0		D					
3										Clay till (as above)	
			3.8	4.0		D					
4											
			4.8	5.0		D					
5											
			5.8	6.0		D					
6											
			6.8	7.0		D			7.00		
7											
8											
9											
10											

Remarks

D = Disturbed sample for FID screening
3m screen w. gravel pack (4-7m bgl)

4m plain tubing + { 1m bentonite plug (3-4m bgl); 2m grout (1-3m bgl); 1m concrete and headworks (0-1m bgl) }

Scale 1:50	Golder Associates	Cable Percussion/Rotary Openhole Record	Project No. 93525199.4000

Project Monkstown		BOREHOLE No. GA5	
Client Northern Telecom	Contractor Stratex	Sheet 1 of 1	
Rig Type Shell & Auger	Location (See Borehole Location Map)	Driller MG	Engineer KF
Hole Diameter 6 1/2 inch	Ground Surface Elevation 36.1 mOD	Date Started 11/03/94	
Completion Depth 8.0m bgl	Casing Details Temporary casing during drilling	Date Completed 11/03/94	

Downhole Depth	Borehole Progress	Depth to Water	Samples/Tests Depth/m From	To	No.	Type	Symbolic Log	Elevation, m OD	Depth, m (thickness m)	Geological Description	Installation Details
									0.05	MADE GROUND - Asphalt	
1			0.9	1.0		D			1.20	MADE GROUND - angular, grey basalt or dolerite cobbles with basaltic sand, gravel and pebbles (becoming wet at 0.9m bgl	
2			1.8	2.0		D				Firm to stiff, red-brown CLAY till with sub angular clasts of sand, gravel and pebble size and occasional cobbles	
3			2.8	3.0		D					
4			3.8	4.0		D			4.60		
									5.00	Soft to firm, red-brown SILT with coarse sand	
5			4.8	5.0		D			5.20	Soft to firm, red-brown SILT	
									5.70	Soft to firm, red-brown, sandy SILT	
6			5.8	6.0		D			6.50	Soft to firm, grey-brown sandy SILT	
7			6.8	7.0		D				Soft, red-brown CLAY with coarse sand and gravel	
8			7.8	8.0		D			8.00		
9											
10											

Remarks
3m screen w. gravel pack (5-8m bgl)
5m plain tubing + { 1m bentonite plug; 3m grout; 1m concrete and headworks }
D = Disturbed sample for FID screening

Scale 1:50	Golder Associates	Cable Percussion/Rotary Openhole Record	Project No. 93525199.4000

Project	Monkstown			BOREHOLE No. GA6
Client	Northern Telecom	Contractor	Stratex	Sheet 1 of 1
Rig Type	Shell & Auger	Location	(See Borehole Location Map)	Driller MG Engineer KF
Hole Diameter	6 1/2 inch	Ground Surface Elevation .	36.5 mOD	Date Started 14/03/94
Completion Depth	8.0m bgl	Casing Details	Temporary casing during drilling	Date Completed 14/03/94

Downhole Depth	Borehole Progress	Depth to Water	Samples/Tests Depth/m From	To	No.	Type	Symbolic Log	Elevation, m OD	Depth, m (thickness m)	Geological Description	Installation Details
									0.20	MADE GROUND - Black, loose ash	
									0.50	Firm, orange-brown silty CLAY	
									0.70	Firm, dark-grey CLAY with abundant fine gravel	
			0.8	1.0		D			0.80	Firm, light brown, mottled CLAY with sand and gravel	
1									1.00	Firm orange-brown, silty CLAY	
									1.50	Soft to firm orange-brown, silty CLAY with gravel	
			1.8	2.0		D			2.00	Soft, orange-brown, fine SAND with clay and silt	
2									2.30	Soft orange-brown, fine SAND with silt and little clay	
			2.8	3.0		D				Firm to stiff, orange-brown CLAY till with sand, gravel, pebbles and occasional cobbles	
3											
			3.8	4.0		D					
4											
			4.8	5.0		D					
5									5.7		
			5.8	6.0		D			6.35	Firm, brown clayey SILT with few pebbles or gravels and some wet fissures	
6									6.70	Very firm, orange-brown CLAY till with angular gravel.	
			6.8	7.0		D				Soft, wet , brown GRAVEL with clay, silt and sand	
7											
			7.8	8.0		D					
8											
9											
10											

Remarks

3m screen w. gravel pack (5-8m bgl)

5m plain tubing +
- 1m bentonite plug (4-5m bgl)
- 3m grout (1-4m bgl)
- 1m concrete and headworks (0-1m bgl)

D = Disturbed sample for FID screening

Scale 1:50	Golder Associates	Cable Percussion/Rotary Openhole Record	Project No. 93525199.4000

Project Monkstown		BOREHOLE No. GA7
Client Northern Telecom	Contractor Stratex	Sheet 1 of 1
Rig Type Shell & Auger	Location (See Borehole Location Map)	Driller MG Engineer KF
Hole Diameter 6 1/2 inch	Ground Surface Elevation 36.1 mOD	Date Started 14/03/94
Completion Depth 10m bgl	Casing Details Temporary casing during drilling	Date Completed 14/03/94

Downhole Depth	Borehole Progress	Depth to Water	Samples/Tests Depth/m From	To	No.	Type	Symbolic Log	Elevation, m OD	Depth, m (thickness m)	Geological Description	Installation Details
									0.10	MADE GROUND - Black, loose ash	
			0.7	0.9		D				MADE GROUND - dense, grey subangular dolerite gravel with sand, pebbles and cobbles of the same material	
1			0.9	1.0		D			0.90		
									1.40	Stiff red-brown CLAY till with sand and gravel	
2			1.8	2.0		D				Firm, red-brown SAND till with clay and silt	
3			2.8	3.0		D			3.10		
4			3.8	4.0		D			4.10	Firm to stiff, red-brown CLAY till with sand and little gravel	
									4.30	Soft, red-brown SAND till with little clay	
5			4.8	5.0		D				Firm, red-brown sandy clay	
6			5.8	6.0		D					
7			6.8	7.0		D			6.90		
8			7.8	8.0		D				Stiff, red-brown CLAY till with sand, gravel and pebbles	
9			8.8	9.0		D					
10			9.8	10.0		D					

Remarks

Borehole backfilled with bentonite from 10m bgl to 7m before setting the screen

D = Disturbed sample for FID screen

3m screen with gravel pack (4-7m bgl)

4m plain tubing with:
- 1m bentonite seal (4-7m bgl)
- 2m grout (1-3m bgl)
- 1m concrete and headworks (0-1m bgl)

Scale 1:50	Golder Associates	Cable Percussion/Rotary Openhole Record	Project No. 93525199.4000

Project Monkstown		BOREHOLE No. GA8
Client Northern Telecom	Contractor Stratex	Sheet 1 of 1
Rig Type Shell & Auger	Location (See Borehole Location Map)	Driller MG Engineer KF
Hole Diameter 6 1/2 inch	Ground Surface Elevation 37.7 mOD	Date Started 14/03/94
Completion Depth 10m bgl	Casing Details Temporary casing during drilling	Date Completed 15/03/94

Downhole Depth	Borehole Progress	Depth to Water	Samples/Tests Depth/m From	To	No.	Type	Symbolic Log	Elevation, m OD	Depth, m (thickness m)	Geological Description	Installation Details
									0.10	MADE GROUND - Asphalt	
									0.60	MADE GROUND - dense, grey, dolerite cobbles and pebbles with some sand	
			0.7	0.9		D					
1									1.60	Soft, green-grey CLAY with occasional flat, elongate shale/slate flakes	
			1.8	2.0		D			1.90	Stiff, grey-brown CLAY with lenses of green sand, gravel and pebbles	
2										Stiff red-brown CLAY till with sand, gravel, pebbles and occasional cobbles	
			2.8	3.0		D					
3											
			3.8	4.0		D					
4											
			4.8	5.0		D					
5											
			5.8	6.0		D					
6											
			6.8	7.0		D					
7											
8											
			8.8	9.0		D					
9											
10			9.8	10.0		D			10.00		

Remarks
D = Disturbed sample for FID screening

3m screen with gravel pack (7-10m bgl)
7m plain tubing with
- 1m bentonite seal (6-7m bgl)
- 5m grout (1-6m bgl)
- 1m concrete and headworks (0-1m bgl)

Scale 1:50	Golder Associates	Cable Percussion/Rotary Openhole Record	Project No. 93525199.4000

Project Monkstown		BOREHOLE No. GA9
Client Northern Telecom	Contractor Stratex	Sheet 1 of 1
Rig Type Shell & Auger	Location (See Borehole Location Map)	Driller CW Engineer KF
Hole Diameter 6 1/2 inch	Ground Surface Elevation 35.8 mOD	Date Started 15/03/94
Completion Depth 7.0m bgl	Casing Details Temporary casing during drilling	Date Completed 15/03/94

Downhole Depth	Borehole Progress	Depth to Water	Samples/Tests Depth/m From	To	No.	Type	Symbolic Log	Elevation, m OD	Depth, m (thickness m)	Geological Description	Installation Details
									0.10	MADE GROUND - Asphalt	
			0.55	0.75		D			0.75	MADE GROUND - dense, dolerite/basalt GRAVEL with angular pebbles and cobbles	
1										Firm to stiff, grey-brown CLAY till with sand and gravel	
			1.8	2.0		D			2.00		
2										Stiff red-brown CLAY till with sand, gravel and pebbles	
3			2.8	3.0		D					
4			3.8	4.0		D					
5			4.8	5.0		D					
6			5.8	6.0		D					
7			6.8	7.0		D			7.00		
8											
9											
10											

Remarks D = Disturbed sample for FID screening

3m screen with gravel pack (4-7m bgl)

4m plain tubing with
- 1m bentonite plug
- 2m grout
- 1m concrete and headworks

Scale 1:50	Golder Associates	Cable Percussion/Rotary Openhole Record	Project No. 93525199.4000

Project Monkstown		BOREHOLE No. GA10
Client Northern Telecom	Contractor Stratex	Sheet 1 of 1
Rig Type Shell & Auger	Location (See Borehole Location Map)	Driller MG Engineer KF
Hole Diameter 6 1/2 inch	Ground Surface Elevation 36.1 mOD	Date Started 15/03/94
Completion Depth 7.0m bgl	Casing Details Temporary casing during drilling	Date Completed 16/03/94

Downhole Depth	Borehole Progress	Depth to Water	Samples/Tests Depth/m From	To	No.	Type	Symbolic Log	Elevation, m OD	Depth, m (thickness m)	Geological Description	Installation Details
									0.60	MADE GROUND- loose, black ash and cinders	
1			0.8	1.0		D			1.70	Firm, grey, sandy CLAY	
2			1.8	2.0		D				Stiff red-brown CLAY till with sand, gravel and pebbles	
3			2.8	3.0		D					
4			3.8	4.0		D					
5			4.8	5.0		D			5.5		
										Firm, red-brown sandy SILT	
									5.8	Firm, red-brown silty CLAY	
6			5.8	6.0		D			6.0	Soft, wet, red-brown sandy GRAVEL with small amounts of clay and silt	
7			6.8	7.0		D			7.0		
8											
9											
10											

Remarks D = Disturbed sample for FID screening

3m screen with gravel pack (4-7m bgl)

4m plain tubing with
- 1m bentonite plug (3-4m bgl)
- 2m grout (1-3m bgl)
- 1m concrete and headworks (0-1m bgl)

Scale 1:50	Golder Associates	Cable Percussion/Rotary Openhole Record	Project No. 93525199.4000

Project Monkstown		BOREHOLE No. GA11
Client Northern Telecom	Contractor Stratex	Sheet 1 of 1
Rig Type Shell & Auger	Location (See Borehole Location Map)	Driller MG Engineer KF
Hole Diameter 6 1/2 inch	Ground Surface Elevation 36.3 mOD	Date Started 15/03/94
Completion Depth 7.0m bgl	Casing Details Temporary casing during drilling	Date Completed 16/03/94

Downhole Depth	Borehole Progress	Depth to Water	Samples/Tests Depth/m From	To	No.	Type	Symbolic Log	Elevation, m OD	Depth, m (thickness m)	Geological Description	Installation Details
									0.05	MADE GROUND - beige, loose gravels	
									0.15	MADE GROUND - asphalt	
1			0.9	1.1		D			1.10	MADE GROUND - loose, grey, dolerite cobbles, gravel and sand	
									1.30	Very stiff, red CLAY till with sand and gravel	
			1.8	2.0		D			1.90	Firm, red-brown SAND with clay and silt	
2									2.80	Soft to firm, red-brown coarse SAND with silt and little clay	
3			2.8	3.0		D					
			3.8	4.0		D			4.00	Stiff, red-brown CLAY till with sand and gravel	
4									4.60	Soft to firm, brown coarse SAND with silt and small amounts of clay and chalk gravel	
									4.90	Firm, red-brown silty SAND with little clay	
5			4.8	5.0		D					
6			5.8	6.0		D					
7			6.8	7.0		D			7.0	Firm, red-brown coarse SAND with much silt and some slightly cemented sandy patches	
8											
9											
10											

Remarks D = Disturbed sample for FID screening 3m screen with gravel pack (4-7m bgl)

4m plain tubing with
- 1m bentonite plug (3-4m bgl)
- 2m grout (1-3m bgl)
- 1m concrete and headworks (0-1m bgl)

Scale 1:50	Golder Associates	Cable Percussion/Rotary Openhole Record	Project No. 93525199.4000

Project Monkstown		BOREHOLE No. GA12
Client Northern Telecom	Contractor Stratex	Sheet 1 of 1
Rig Type Shell & Auger	Location (See Borehole Location Map)	Driller MG Engineer KF
Hole Diameter 6 1/2 inch	Ground Surface Elevation 36.3 mOD	Date Started 21/3/94
Completion Depth 8.5m bgl	Casing Details Temporary casing during drilling	Date Completed 22/3/94

Downhole Depth	Borehole Progress	Depth to Water	Samples/Tests Depth/m From	To	No.	Type	Symbolic Log	Elevation, m OD	Depth, m (thickness m)	Geological Description	Installation Details
									0.60	MADE GROUND Paving stones, sand, gravel and cobble fill. FILL becomes finer with depth	
1			0.8	1.0		D				Stiff, red-brown CLAY with subrounded gravel and pebbles	
2			1.8	2.0		D					
3			2.8	3.0		D					
4			3.8	4.0		D					
5			4.8	5.0		D			5.80		
6			5.8	6.0		D			6.15	Stiff, red-brown CLAY till with much silt and fine sand and few gravels or pebbles	
										Firm, red-brown silty CLAY with sand, gravel and pebbles	
7			6.8	7.0		D			7.00		
8			7.8	8.0		D			8.50	Soft, wet red-brown coarse SAND and fine gravel with occasional pebbles and little clay	
9											
10											

Remarks D = Disturbed sample for FID screening

3m screen with gravel pack (5.5-8.5m bgl)

5.5m plain tubing with
- 1m bentonite (4.5-5.5m bgl)
- 3.5m grout (1-4.5m bgl)
- 1m concrete and headworks (0-1m bgl)

Scale 1:50	Golder Associates	Cable Percussion/Rotary Openhole Record	Project No. 93525199.4000

Project Monkstown		BOREHOLE No. GA13
Client Northern Telecom	Contractor Stratex	Sheet 1 of 1
Rig Type Shell & Auger	Location (See Borehole Location Map)	Driller MG Engineer KF
Hole Diameter 6 1/2 inch	Ground Surface Elevation 36.3 mOD	Date Started 22/3/94
Completion Depth 8.0m bgl	Casing Details Temporary casing during drilling	Date Completed 23/3/94

Downhole Depth	Borehole Progress	Depth to Water	Samples/Tests Depth/m From	To	No.	Type	Symbolic Log	Elevation, m OD	Depth, m (thickness m)	Geological Description	Installation Details
									0.25	MADE GROUND - paving, concrete, sand and gravel fill	
1			0.8	1.0		D				Stiff, dry, red-brown CLAY till with some sand, gravel, pebbles and occasional cobbles	
2			1.8	2.0		D			2.70		
3			2.8	3.0		D				Firm, red-brown coarse SAND and fine gravel with silt and clay and occasional cobbles	
4			3.8	4.0		D					
5			4.8	5.0		D					
6			5.8	6.0		D			6.50		
7			6.8	7.0		D			6.80	Soft, wet, red-brown coarse SAND and fine gravel with silt and clay lenses	
									7.50	Soft, wet, grey-brown coarse SAND and fine gravel with silt and clay lenses	
									7.90	Stiff, red-brown CLAY till with sand, gravel and pebbles	
8			7.8	8.0		D			8.00	Firm, red-brown clayey SAND	
9											
10											

Remarks D = Disturbed sample for FID screening 3m screen with gravel pack (5-8m bgl)

5m plain tubing with
- 1m bentonite (5-8m bgl)
- 3m grout (1-4m bgl)
- 1m concrete and headworks (0-1m bgl)

Scale 1:50	Golder Associates	Cable Percussion/Rotary Openhole Record	Project No. 93525199.4000

Project Monkstown		BOREHOLE No. GA14
Client Northern Telecom	Contractor Stratex	Sheet 1 of 1
Rig Type Shell & Auger	Location (See Borehole Location Map)	Driller MG Engineer KF
Hole Diameter 6 1/2 inch	Ground Surface Elevation 36.1 mOD	Date Started 22/3/94
Completion Depth 8.0m bgl	Casing Details Temporary casing during drilling	Date Completed 24/3/94

Downhole Depth	Borehole Progress	Depth to Water	Samples/Tests Depth/m From	To	No.	Type	Symbolic Log	Elevation, m OD	Depth, m (thickness m)	Geological Description	Installation Details
									0.20	MADE GROUND - Asphalt	
									0.75	MADE GROUND - dense dolerite gravel with angular cobbles, pebbles and sand. Becoming finer with depth	
			0.8	1.0		D					
1										Stiff, red-brown CLAY till with sand, gravel, pebbles and occasional subrounded cobbles	
			1.8	2.0		D					
2											
			2.8	3.0		D					
3											
			3.8	4.0		D					
4											
			4.8	5.0		D					
5											
			5.8	6.0		D					
6											
			6.8	7.0		D					
7											
			7.8	8.0		D			8.00		
8											
9											
10											

Remarks D = Disturbed sample for FID screening 3m screen with gravel pack (5-8m bgl)

5m plain tubing with { 1m bentonite (5-8m bgl)
3.5m grout (1-4m bgl)
1m concrete and headworks (0-1m bgl) }

Scale 1:50	Golder Associates	Cable Percussion/Rotary Openhole Record	Project No. 93525199.4000

Project Remediation/Monkstown		BOREHOLE No. GA16
Client Northern Telecom	Contractor Glover	Sheet 1 of 2
Rig Type Shell & Auger	Location (See Borehole Location Map)	Driller JC Engineer KF
Hole Diameter 8 1/2 inch	Ground Surface Elevation 36.09m aOD	Date Started 15/08/94
Completion Depth 11.5m bgl	Final Depth 11.5m bgl	Date Completed 16/08/94

Downhole Depth	Borehole Progress	Depth to Water	Samples/Tests Depth/m From	To	No.	Type	Symbolic Log	Elevation, m OD	Depth, m (thickness m)	Geological Description	Installation Details
								0.15		MADE GROUND - Asphalt	
										MADE GROUND - fill material consisting of sand and gravel and cobbles of dolerite	
1											
			1.3	1.3		FIM 15 ppm		1.3		Grey clay containing organic remains and sand and gravel	
			1.6	1.6		FIM 1 ppm		1.6		Soft to firm, red/brown CLAY containing sand and gravel sized particles and occasional pebbles. Disturbed, soft to firm (Driving large cobble from 1.6 to 5.7 so ALL samples are disturbed, no SPT or U100 possible)	
2											
			2.5	2.5		FIM 0 ppm					
3											
			3.5	3.5		FIM 0 ppm					
4											
			4.5	4.5		FIM 1.5 ppm					
5											
			5.5	5.5		FIM 1.5 ppm		5.7		Stiff red-brown CLAY containing sand, gravel and pebbles	
			5.6	5.6		FIM 0 ppm					
6			6.0	6.0		FIM 0.5 ppm					
7											
8			8.0	8.0		FIM 0 ppm					
			8.5	8.95		U100 (GA16 U1)					
9			9.00			SPT (40 blows/ foot)					
						SPT (57 blows/ foot)					
10											

Remarks 3m screen with gravel pack (8.5 - 11.5m bgl)
8.5m plain tubing with
- 1m bentonite plug (7.5 - 8.5m bgl)
- 6.5m grout (1 - 7.5m bgl)
- 1m concrete and headworks (0 - 1m bgl)

Scale 1:50	Golder Associates	Cable Percussion/Rotary Openhole Record	Project No. 94525110

Project Remediation/Monkstown		BOREHOLE No. GA16
Client Northern Telecom	Contractor Glover	Sheet 2 of 2
Rig Type Shell & Auger Hole Diameter 8 1/2 inch Completion Depth 11.5m bgl	Location (See Borehole Location Map) Ground Surface Elevation 36.09m aOD Final Depth 11.5m bgl	Driller JC Engineer KF Date Started 15/08/94 Date Completed 16/08/94

Downhole Depth	Borehole Progress	Depth to Water	Samples/Tests Depth/m From	To	No.	Type	Symbolic Log	Elevation, m OD	Depth, m (thickness m)	Geological Description	Installation Details
										(See sheet 1 of 2)	
11											
								11.5		Base of Borehole	
12											

Remarks

Scale 1:50	Golder Associates	Cable Percussion/Rotary Openhole Record	Project No. 94525110

Project Remediation/Monkstown		BOREHOLE No. GA17
Client Northern Telecom	Contractor Glover	Sheet 1 of 2
Rig Type Shell & Auger	Location (See Borehole Location Map)	Driller JC Engineer KF
Hole Diameter 8 1/2 inch	Ground Surface Elevation 36.07	Date Started 15/08/94
Completion Depth 8.4m bgl	Final Depth 12m bgl	Date Completed 16/08/94

Downhole Depth	Borehole Progress	Depth to Water	Samples/Tests Depth/m From	To	No.	Type	Symbolic Log	Elevation, m OD	Depth, m (thickness m)	Geological Description	Installation Details
								0.15		MADE GROUND - Asphalt	
										MADE GROUND - Fill of dense grey, some gravel and angular cobbles (mainly dolerite)	
1			1.0	1.0		FIM 0 ppm		1.0		Stiff red/brown CLAY containing sand and gravel sized particles	
2			2.0	2.0		FIM 1.5ppm		2.0		Friable red/brown CLAY with sand and gravel	
								2.75		Brown coarse SAND and gravel	
3			3.0	3.0		FIM 1 ppm				Soft to firm red/brown clayey SILT	
4			4.0	4.0		FIM 0.5 ppm					
5			5.0	5.0		FIM 10 ppm		4.9		Brown, friable, wet medium grained SAND containing silt and a little clay (No gravel or pebbles)	
6			6.0	6.0		FIM 25 ppm		6.0		Stiff dry, red/brown clayey SILT	
								6.1		Stiff, brown CLAY containing occasional fine gravel sized particles	
								6.7		Dark brown, soft, clayey SAND	
7			7.0	7.0		FIM 20 ppm					
8			8.0	8.0		FIM 1.6 ppm		8.0		Silty SAND red-brown (firm but crumbly)	
								8.1		Red brown soft to firm clayey SILT	
								8.8		Red brown stiff CLAY with sand, gravel and pebble sized clasts	
9			9.0	9.0		FIM 1.0 ppm					
10			10.0	10.0		FIM 0 ppm					

Remarks Hole backfilled with bentonite from 12m bgl to 8.4m bgl
3m screen with gravel pack (5.4-8.4m bgl)
5.4m plain tubing with
1m bentonite plug (4.4 - 5.4m bgl)
3.4m grout (1.0 - 4.4m bgl)
1m concrete and headworks (0 -1m bgl)

Scale 1:50	Golder Associates	Cable Percussion/Rotary Openhole Record	Project No. 94525110

Project	Remediation/Monkstown			BOREHOLE No. GA17
Client	Northern Telecom	Contractor	Glover	Sheet 2 of 2
Rig Type	Shell & Auger	Location	(See Borehole Location Map)	Driller JC Engineer KF
Hole Diameter	8 1/2 inch	Ground Surface Elevation	36.07	Date Started 15/08/94
Completion Depth	8.4m bgl	Final Depth	12m bgl	Date Completed 16/08/94

Downhole Depth	Borehole Progress	Depth to Water	Samples/Tests Depth/m From	To	No.	Type	Symbolic Log	Elevation, m OD	Depth, m (thickness m)	Geological Description	Installation Details
11			11.0	11.0		FIM 0.5ppm					
12			12.0	12.0		FIM 0 ppm		12.0		Base of Borehole	

Remarks

Scale 1:50	Golder Associates	Cable Percussion/Rotary Openhole Record	Project No. 94525110

Project	Remediation/Monkstown			BOREHOLE No. GA18
Client	Northern Telecom	Contractor	Glover	Sheet 1 of 2
Rig Type	Shell & Auger	Location	(See Borehole Location Map)	Driller JC Engineer KF
Hole Diameter	8 1/2 inch	Ground Surface Elevation	36.15m aOD	Date Started 18/8/94
Completion Depth	Multi-level piezometer	Final Depth	11.1m bgl	Date Completed 18/8/94

Downhole Depth	Borehole Progress	Depth to Water	Samples/Tests Depth/m From	To	No.	Type	Symbolic Log	Elevation, m OD	Depth, m (thickness m)	Geological Description	Installation Details
										MADE GROUND - Asphalt	
								0.15		MADE GROUND - Fill material containing dense sand, gravel, pebbles and cobbles of dolerite	
1			1.0	1.0		FIM 0.5 ppm		1.0		Wet red/brown CLAY with sand, gravel and pebbles	
								1.2		Wet red brown clayey SAND, soft medium grained (becoming coarser grained towards base)	
2			2.0	2.0		FIM 9 ppm		2.2		Red brown stiff pebbly CLAY	
								2.3		Dry crumbly brown coarse grained SAND and gravel with silt, clay and pebbles (some slightly cemented patches)	
			2.5	2.5		SPT (18 blows)					
3			3.0	3.0		PSD					
			3.0	3.0		FIM 5 ppm					
			3.8	3.8		FIM 5 ppm					
4			4.2	4.2		FIM 4 ppm		4.2		Firm, crumbly, red brown clayey SILT	
								4.5		Soft to firm red brown sandy SILT (medium to fine grained sand) Occasional pebbles (rounded) becoming damp at 5.5m bgl	
5			5.5	5.5		FIM 1 ppm		5.7		Stiff red brown silty CLAY	
								5.8		Soft to firm red-brown clayey SILT containing sand. (increasing clay content with depth)	
6			6.7	6.7		FIM 6.5 ppm					
7			7.1	7.1		SPT 30 blow					
			7.1	7.1		FIM 0 ppm					
8			8.1	8.1		FIM 0.5 ppm					
								8.7		Stiff red brown silty CLAY	
9											
10											

Remarks Hole backfilled with bentonite from 11.1m bgl to 8.8m bgl
Lower Piezometer (GA18B) : 0.3m piezometer tip (8.4 - 8.7m bgl with sand pack (8.3 - 8.8m bgl)
8.4m plain tubing with bentonite plug (5.7 - 8.3m bgl)
Upper Piezometer (GA18A): 0.3m piezometer tip (5.3 - 5.6m bgl) with sand pack (5.7 - 5.2m bgl)
5.3m plain tubing with - bentonite plug (4.2 - 5.2m bgl, grout (1 - 4.2m bgl), concrete and headworks (0-1m bgl)

Scale 1:50	Golder Associates	Cable Percussion/Rotary Openhole Record	Project No. 94525110

Project Remediation/Monkstown		BOREHOLE No. GA18
Client Northern Telecom	Contractor Glover	Sheet 2 of 2
Rig Type: Shell & Auger	Location: (See Borehole Location Map)	Driller: JC / Engineer: KF
Hole Diameter: 8 1/2 inch	Ground Surface Elevation: 36.15m aOD	Date Started: 18/8/94
Completion Depth: Multi-level piezometer	Final Depth: 11.1m bgl	Date Completed: 18/8/94

Downhole Depth	Borehole Progress	Depth to Water	Samples/Tests Depth/m From	To	No.	Type	Symbolic Log	Elevation, m OD	Depth, m (thickness m)	Geological Description	Installation Details
										Stiff red brown silty CLAY	
11								11.1		Base of Borehole	
12											

Remarks

Scale 1:50	Golder Associates	Cable Percussion/Rotary Openhole Record	Project No. 94525110

Project	Remediation/Monkstown			BOREHOLE No. GA19
Client	Northern Telecom	Contractor	Glover	Sheet 1 of 2
Rig Type	Shell & Auger	Location	(See Borehole Location Map)	Driller JC Engineer KF
Hole Diameter	8 1/2 inch	Ground Surface Elevation	36.07m bgl	Date Started 18/8/94
Completion Depth	Multi level standpipe	Final Depth	11.55m bgl	Date Completed 23/8/94

Samples/Tests

From	To	No.	Type
2.0	2.0		FIM 5ppm
2.5	2.95		SPT 18 blow
			FIM 15ppm
4.5	4.5		FIM 30ppm
5.0	5.0		FIM 800ppm
5.5	5.5		FIM 800ppm
5.9	5.9		FIM 500
6.0	6.45		SPT 47 blows
6.0	6.45		FIM 300 ppm
7.0	7.0		FIM 7ppm
7.7	7.7		Soil VOC
8.0	8.0		FIM 2ppm
8.7	9.0		PSD
8.7	9.0		FIM 2ppm
9.4	9.85		SPT 29 blows
9.4	9.85		FIM 2ppm

Geological Description

Elevation, m OD	Geological Description
	MADE GROUND - Asphalt
0.15	MADE GROUND - fill material of gravel, pebbles and ashy material
0.9	Stiff, red/brown CLAY with silt sand and gravel
1.3	Firm, crumbly, red/brown clayey SILT
2.0	Stiff red/brown CLAY containing sand, gravel and pebbles (Abundant chalk pebbles)
2.2	Firm crumbly red/brown SILT containing fine sand
2.8	Firm crumbly red/brown clay SILT
3.9	Firm red/brown sandy SILT, crumbles easily (strong smell of solvents) containing occasional gravel and pebbles (mainly of chalk and flint)
5.7	Soft wet brown coarse SAND with white angular flint gravel flakes
5.9	Stiff yellow-brown sandy SILT
6.3	Soft wet red/brown silty SAND
7.8	Stiff brown CLAY with sand, gravel and pebbles
8.0	Firm red brown SILT with clay and occasional gravel sized particles

Remarks Hole backfilled with bentonite and gravel from 11.55m bgl to 9m bgl
Lower standpipe (GA19B): 1m screen with gravel pack (8-9m bgl) 8m plain tubing with bentonite plug (6.6-8m bgl)
Upper standpipe (GA19A): 1m screen with gravel pack (5.6 - 6.6m bgl)
5.6m plain tubing with - 1m bentonite plug (4.6-5.6m bgl, 3.6m grout (1 - 4.6m bgl), 1m concrete and headworks (0-1m bgl)

Scale 1:50	Golder Associates	Cable Percussion/Rotary Openhole Record	Project No. 94525110

Project Remediation/Monkstown		BOREHOLE No. GA19
Client Northern Telecom	Contractor Glover	Sheet 2 of 2
Rig Type Shell & Auger	Location (See Borehole Location Map)	Driller JC Engineer KF
Hole Diameter 8 1/2 inch	Ground Surface Elevation 36.07m bgl	Date Started 18/8/94
Completion Depth Multi level standpipe	Final Depth 11.55m bgl	Date Completed 23/8/94

Downhole Depth	Borehole Progress	Depth to Water	Samples/Tests Depth/m From	To	No.	Type	Symbolic Log	Elevation, m OD	Depth, m (thickness m)	Geological Description	Installation Details
								10.0		Stiff red/brown CLAY with sand, gravel and pebbles, and occasional cobbles	
			10.9	11.55		U100					
11											
								11.55		BASE OF BOREHOLE	
12											
13											

Remarks

Scale 1:50	Golder Associates	Cable Percussion/Rotary Openhole Record	Project No. 94525110

Project Remediation/Monkstown		BOREHOLE No. GA20
Client Northern Telecom	Contractor Glover	Sheet 1 of 2
Rig Type Shell & Auger	Location (See Borehole Location Map)	Driller JC Engineer KF
Hole Diameter 8 1/2 inch	Ground Surface Elevation 35.90m aOD	Date Started 19/8/94
Completion Depth	Final Depth 18.5m bgl	Date Completed 23/8/94

Downhole Depth	Borehole Progress	Depth to Water	Samples/Tests Depth/m From	To	No.	Type	Symbolic Log	Elevation, m OD	Depth, m (thickness m)	Geological Description	Installation Details
								0.15		MADE GROUND - Asphalt	
										MADE GROUND - Fill material of dense sand, gravel, pebbles and cobbles	
1			1.2	1.2		FIM 0ppm		1.0		Stiff red/brown CLAY with silt, sand, gravel and pebbles	
2			2.0	2.0		FIM 0.5ppm					
3			3.0	3.0		FIM 0.5ppm					
			3.2	3.2		FIM 0.5ppm		3.2		Soft to firm red/brown sandy SILT	
4			4.2	4.2		FIM 0ppm					
			4.6	4.6		FIM 0ppm		4.4		Soft to firm friable red/brown SILT with sand and abundant gravel, occasional large chalk cobble	
5			5.1	5.1		FIM 0.5ppm		5.0		Soft to firm, friable, red brown SILT with fine sand. No coarser material	
			5.7	5.7		Soil VOC		5.7		Soft to firm, friable, red-brown clayey SILT	
6			6.0	6.0		GASTEC					
			6.2	6.2		FIM 1ppm					
			6.6	7.05		SPT 35 blows					
			6.8	6.8		FIM 0.5ppm					
7			7.7	7.7		FIM 0ppm		7.32		Soft brown wet silty CLAY with occasional gravel and pebbles	
8											
9			9.1	9.1		FIM 0ppm		9.1		Stiff red brown CLAY with sand gravel and pebbles and occasional pebbles	
			9.2	9.65		SPT 30 blows					
10											

Remarks Hole backfilled with gravel and bentonite from 18.5m bgl to 15.2m bgl
Lower piezometer (GA20B): 0.3m piezometer screen (14.7 - 15m bgl() with sand pack (14.0 - 15.2m bgl)
14.7m of plain tubing with - 1m bentonite plug (13 -14m bgl), 2.8m grout (8.2 - 11m bgl)
2m gravel (11 - 13m bgl), 1m bentonite plug (8.2 - 7.2m bgl

Scale 1:50	Golder Associates	Cable Percussion/Rotary Openhole Record	Project No. 94525110

Project Remediation/Monkstown		BOREHOLE No. GA20
Client Northern Telecom	Contractor Glover	Sheet 2 of 2
Rig Type Shell & Auger	Location (See Borehole Location Map)	Driller JL Engineer KF
Hole Diameter 8 1/2 inch	Ground Surface Elevation 35.90m aOD	Date Started 19/8/94
Completion Depth	Final Depth 18.5m bgl	Date Completed 23/8/94

Downhole Depth	Borehole Progress	Depth to Water	Samples/Tests Depth/m From	To	No.	Type	Symbolic Log	Elevation, m OD	Depth, m (thickness m)	Geological Description	Installation Details
			10.1	10.1		FIM 0ppm				Stiff red-brown CLAY etc (as sheet 1 of 2)	
11			11.1	11.1		FIM 0.5ppm					
12			12.1	12.1		FIM 0ppm					
13			13.1	13.1		FIM 0ppm		13.4			
14										Soft to firm brown friable sandy CLAY with abundant gravel and pebble sized particles (angular flint, chalk and basalt) wet	
			14.5	14.5		FIM 0.5ppm		14.5			
15										Stiff red brown sandy SILT with semi-rounded gravel, pebbles and occasonal cobbles of basalt or green sandstone	
			15.5	15.5		FIM 0ppm					
16			16.5	16.5		FIM 1ppm					
17											
18								18.2		Stiff, saturated, red-brown silty CLAY with sand, gravel, pebbles and cobbles	
								18.5		BASE OF BOREHOLE 18.5m	
19											
20											

Remarks

Upper piezometer (GA20A): 0.3m piezometer screen (6.7 - 7.0m bgl) with sand pack (6.0 - 7.2m bgl)
6.7m of plain tubing with:
- 1m bentonite plug (5 - 6m bgl)
- 4m grout (1-5m bgl)
- 1m concrete and headworks (0-1m bgl)

Scale 1:50	Golder Associates	Cable Percussion/Rotary Openhole Record	Project No. 94525110

Project	Remediation/Monkstown			BOREHOLE No. GA21	
Client	Northern Telecom	Contractor	Glover	Sheet 1 of 1	
Rig Type	Shell & Auger	Location	(See Borehole Location Map)	Driller JC	Engineer KF
Hole Diameter	8 1/2 inch	Ground Surface Elevation	36.01m aOD	Date Started	15/08/94
Completion Depth	7.5m bgl	Final Depth	10.3m bgl	Date Completed	16/08/94

Downhole Depth	Borehole Progress	Depth to Water	Samples/Tests Depth/m From	To	No.	Type	Symbolic Log	Elevation, m OD	Depth, m (thickness m)	Geological Description	Installation Details
								0.1		MADE GROUND - Asphalt	
										MADE GROUND - Fill material containing dense, grey sand, gravel, pebbles and cobbles	
1			1.0	1.0		FIM 2.5ppm		1.15		Firm red brown silty CLAY with occasional sand and gravel	
								1.50		Medium brown fine sandy SILT with some clay	
2			2.0	2.0		FIM 2ppm					
			2.7	3.15		SPT 11 blows					
			2.7	2.9		VOC FIM 2ppm		2.80		Medium brown very silty fine SAND with occasional thin horizon of silty clay	
3			3.0	3.0		FIM 2ppm		3.00		Firm brown silty fine SAND	
								3.50		Stiff brown clayey SILT lense - then similar material as above	
4			4.0	4.0		FIM 7ppm		4.2		Soft brown fine silty SAND (damp)	
								4.6		Soft wet clayey, sandy SILT	
5			5.0	5.0		FIM 8ppm					
								5.60		Layer of stiff brown silty CLAY with much gravel and occasional cobbles of chalk	
			5.8	5.8		FIM 7ppm		5.80		Soft, wet, brown, fine to medium grained SAND	
6			6.0	6.0		FIM 9ppm					
7											
			7.50	7.50		FIM 0ppm		7.5		Stiff, saturated, red brown CLAY with sand, gravel and pebbles	
8											
			8.50	8.50		FIM 0ppm					
9								9.2		Firm to stif,f dark brown, clayey SILT - damp but not saturated	
10								10.3		End of hole 10.3m	

Remarks Hole backfilled from 10.3m bgl to 7.5m bgl
2m screen with gravel pack (5.5 - 7.5m bgl)
5.5m plain tubing with
- 1.5m bentonite plug (4 - 5.5m bgl)
- 3m grout (1 - 3m bgl)
- 1m concrete and headworks (0 - 1m bgl)

Scale 1:50	Golder Associates	Cable Percussion/Rotary Openhole Record	Project No. 94525110

APPENDIX 2 - DETAILED GROUNDWATER CHEMISTRY

Source: Golder Associates (2001)

ORGANIC PARAMETERS

DOWNGRADIENT

BH	Date	TCE µg/L	c-DCE µg/L	VC µg/L	TCA µg/L
MWD	18/4/96	12,000	150	ND	ND
MWD	17/7/96	12,000	430	ND	ND
MWD	24/10/96	9,300	270	ND	ND
MWD	21/1/97	7,400	400	ND	ND
MWD	12/8/97	3300	3000	ND	ND
MWD	4/4/98	2400	3700	ND	ND
MWD	29/1/99	23	4099	ND	ND
MWD	18/7/00	440	710	ND	ND
MWD	28/2/01	NS	NS	NS	NS
GA12	19/3/94	4,900	NA	ND	ND
GA12	26/8/94	6,500	NA	ND	ND
GA12	18/4/96	3,000	120	ND	ND
GA12	17/7/96	2,700	180	ND	ND
GA12	24/10/96	4,300	260	ND	ND
GA12	21/1/97	1,400	88	ND	ND
GA12	12/8/97	1,200	86	ND	ND
GA12	4/4/98	800	71	ND	ND
GA12	29/1/99	212	14	ND	ND
GA12	18/7/00	1,100	130	ND	ND
GA12	28/2/01	540	46	ND	ND
GA13	19/3/94	43,000	NA	ND	ND
GA13	26/8/94	28,000	NA	ND	ND
GA13	18/4/96	260,000	ND	ND	ND
GA13	17/7/96	150,000	430	ND	ND
GA13(A)	24/10/96	85,000	340	ND	ND
GA13(A)	24/10/96	100,000	460	ND	ND
GA13	21/1/97	97,000	290	ND	ND
GA13	12/8/97	130,000	730	ND	ND
GA13	4/4/98	45,000	390	ND	ND
GA13	29/1/99	37,400	1524	ND	ND
GA13	18/7/00	NS	NS	NS	NS
GA13	28/2/01	61,000	1200	ND	ND
GA14	19/3/94	NA	NA	ND	ND
GA14	26/8/94	NA	NA	ND	ND
GA14	18/4/96	1	ND	ND	ND
GA14	17/7/96	2	ND	ND	ND
GA14	24/10/96	1	ND	ND	ND
GA14	21/1/97	1	ND	ND	ND
GA14(dup)	21/1/97	2	ND	ND	ND
GA14	4/4/98	0.7	ND	ND	ND
GA14	29/1/99	ND	ND	ND	ND
GA14	18/7/00	ND	ND	ND	ND
GA14	28/2/01	Hole dry			

NS = location Not Sampled

ND = parameter Not Detected

NA = parameter Not Analysed

ORGANIC PARAMETERS

UPGRADIENT - continued

BH	Date	TCE µg/L	c-DCE µg/L	VC µg/L	TCA µg/L
GA7	19/3/94	30,000	NA	NA	NA
GA7	26/8/94	49,900	NA	NA	NA
GA7	18/4/96	130,000	700	ND	ND
GA7	17/7/96	57,000	340	ND	ND
GA7(dup)	24/10/96	90,000	600	ND	ND
GA7	24/10/96	88,000	560	ND	ND
GA7	21/1/97	27,000	290	ND	ND
GA7(dup)	22/2/97	39,000	240	ND	ND
GA7	12/8/97	21,000	ND	ND	ND
GA7	4/4/98	18,000	670	ND	ND
GA7	29/1/99	15,200	570	ND	ND
GA7	18/7/00	9,800	150	ND	ND
GA7	28/2/01	2,800	50	ND	ND
BH19	19/3/94	250,000	NA	NA	95
BH19	26/8/94	250,000	NA	NA	NA
BH19	18/4/96	95,000	450	ND	ND
BH19	17/7/96	N S	N S	N S	N S
BH19	24/10/96	39,000	240	ND	ND
BH19	21/1/97	14,000	670	ND	ND
BH19	12/8/97	49,000	550	ND	ND
BH19	4/4/98	22,000	290	ND	ND
BH19	29/1/99	11,000	261	ND	ND
BH19 (dup)	29/1/99	13,000	297	ND	ND
BH19	18/7/00	12,000	230	ND	ND
BH19	28/2/01	4,000	130	ND	ND
MWU	18/4/96	79,000	64	ND	ND
MWU	17/7/96	30,000	270	ND	ND
MWU	24/10/96	47,000	2,400	ND	ND
MWU	21/1/97	62,000	5,600	ND	ND
MWU	12/8/97	21,000	ND	ND	ND
MWU	4/4/98	18,000	6,200	530	ND
MWU	29/1/99	9,630	1,454	ND	ND
MWU	18/7/00	510	360	ND	ND
MWU	28/2/01	660	150	ND	ND

NS = location Not Sampled

ND = parameter Not Detected

NA = parameter Not Analysed

ORGANIC PARAMETERS

WITHIN REACTOR

BH	Date	TCE µg/L	c-DCE µg/L	VC µg/L	TCA µg/L
R5	18/4/96	38,000	57	ND	ND
R5	17/7/96	20,000	420	ND	ND
R5	24/10/96	9,400	380	ND	ND
R5	21/1/97	12,000	550	ND	ND
R5	12/8/97	8100	7,200	ND	ND
R5	4/4/98	6,700	3,300	ND	ND
R5	29/1/99	5,600	459	ND	ND
R5	18/7/00	180	190	ND	ND
R5	28/2/01	450	57	ND	ND
R4	18/4/96	14	2	0.4	ND
R4	17/7/96	44	8	ND	ND
R4	24/10/96	100	9	ND	ND
R4 (dup)	24/10/96	100	8	ND	ND
R4	21/1/97	96	8	ND	ND
R4	12/8/97	7	39	2	ND
R4	4/4/98	33	81	2	ND
R4	29/1/99	12	362	ND	ND
R4	18/7/00	3	1	ND	ND
R4	28/2/01	0.9	1	ND	ND
R3	18/4/96	16	ND	ND	ND
R3	17/7/96	5	ND	ND	ND
R3	24/10/96	27	2	ND	ND
R3	21/1/97	33	2	ND	ND
R3	12/8/97	7	13	ND	ND
R3	4/4/98	1	4	ND	ND
R3	29/1/99	ND	ND	ND	ND
R3	18/7/00	ND	ND	ND	ND
R3	28/2/01	ND	ND	ND	ND
R2	18/4/96	73	0.8	ND	ND
R2	17/7/96	11	0.3	ND	ND
R2	24/10/96	14	1	ND	ND
R2	21/1/97	24	1	ND	ND
R2	12/8/97	1	0.6	ND	ND
R2	4/4/98	9	6	ND	ND
R2	29/1/99	ND	ND	ND	ND
R2	18/7/00	Sample	Broken		
R2	28/2/01	ND	ND	ND	ND
R1	18/4/96	25	0.8	ND	ND
R1	17/7/96	17	0.5	ND	ND
R1	24/10/96	20	2	ND	ND
R1	21/1/97	2	0.1	ND	ND
R1(dup)	22/1/97	2	0.1	ND	ND
R1	12/8/97	ND	ND	ND	ND
R1	4/4/98	3	2	ND	ND
R1	29/1/99	ND	ND	ND	ND
R1	18/7/00	ND	ND	ND	ND
R1	28/2/01	ND	ND	ND	ND

NS = location Not Sampled
ND = parameter Not Detected
NA = parameter Not Analysed

ORGANIC PARAMETERS

DOWNGRADIENT

BH	Date	TCE µg/L	c-DCE µg/L	VC µg/L	TCA µg/L
MWD	18/4/96	12,000	150	ND	ND
MWD	17/7/96	12,000	430	ND	ND
MWD	24/10/96	9,300	270	ND	ND
MWD	21/1/97	7,400	400	ND	ND
MWD	12/8/97	3300	3000	ND	ND
MWD	4/4/98	2400	3700	ND	ND
MWD	29/1/99	23	4099	ND	ND
MWD	18/7/00	440	710	ND	ND
MWD	28/2/01	NS	NS	NS	NS
GA12	19/3/94	4,900	NA	ND	ND
GA12	26/8/94	6,500	NA	ND	ND
GA12	18/4/96	3,000	120	ND	ND
GA12	17/7/96	2,700	180	ND	ND
GA12	24/10/96	4,300	260	ND	ND
GA12	21/1/97	1,400	88	ND	ND
GA12	12/8/97	1,200	86	ND	ND
GA12	4/4/98	800	71	ND	ND
GA12	29/1/99	212	14	ND	ND
GA12	18/7/00	1,100	130	ND	ND
GA12	28/2/01	540	46	ND	ND
GA13	19/3/94	43,000	NA	ND	ND
GA13	26/8/94	28,000	NA	ND	ND
GA13	18/4/96	260,000	ND	ND	ND
GA13	17/7/96	150,000	430	ND	ND
GA13(A)	24/10/96	85,000	340	ND	ND
GA13(A)	24/10/96	100,000	460	ND	ND
GA13	21/1/97	97,000	290	ND	ND
GA13	12/8/97	130,000	730	ND	ND
GA13	4/4/98	45,000	390	ND	ND
GA13	29/1/99	37,400	1524	ND	ND
GA13	18/7/00	NS	NS	NS	NS
GA13	28/2/01	61,000	1200	ND	ND
GA14	19/3/94	NA	NA	ND	ND
GA14	26/8/94	NA	NA	ND	ND
GA14	18/4/96	1	ND	ND	ND
GA14	17/7/96	2	ND	ND	ND
GA14	24/10/96	1	ND	ND	ND
GA14	21/1/97	1	ND	ND	ND
GA14(dup)	21/1/97	2	ND	ND	ND
GA14	4/4/98	0.7	ND	ND	ND
GA14	29/1/99	ND	ND	ND	ND
GA14	18/7/00	ND	ND	ND	ND
GA14	28/2/01	Hole dry			

NS = location Not Sampled

ND = parameter Not Detected

NA = parameter Not Analysed

ORGANIC PARAMETERS

DOWNGRADIENT - continued

BH	Date	TCE µg/L	c-DCE µg/L	VC µg/L	TCA µg/L
GA6	19/3/94	15,000	NA	ND	ND
GA6	26/8/94	7,200	NA	ND	ND
GA6	18/4/96	14,000	990	ND	ND
GA6	17/7/96	8,000	760	ND	ND
GA6	24/10/96	7,600	3200	ND	ND
GA6	21/1/97	4,600	2500	ND	ND
GA6	12/8/97	1,100	9500	ND	ND
GA6	4/4/98	1,400	4000	ND	ND
GA6	29/1/99	NS	NS	NS	NS
GA6	18/7/00	NS	NS	NS	NS
GA6	28/2/01	3,800	1,600	ND	ND
GA10	26/8/94	4,300	NA	ND	ND
GA10	18/4/96	8,700	200	ND	ND
GA10	17/7/96	3,800	150	ND	ND
GA10	24/10/96	4,100	930	ND	ND
GA10	21/1/97	1700	280	ND	ND
GA10	12/8/97	1500	230	ND	ND
GA10	4/4/98	2400	570	ND	ND
GA10	29/1/99	NS	NS	NS	NS
GA10	18/7/00	NS	NS	NS	NS
GA10	28/2/01	Could not locate hole			

NS = location Not Sampled

ND = parameter Not Detected

NA = parameter Not Analysed

INORGANIC PARAMETERS

UPGRADIENT

BH	Date	pH	Chloride mg/L	Sulphate mg/L	Alkalinity mg/L	Ca mg/L	Fe mg/L	Mg mg/L	Mn mg/L	Sulphide mg/L	K mg/L	Na mg/L	S (free) mg/L	Nitrate mg/L
GA5	4/4/98	7.72	35	38	200	37.9	ND	16	ND	0.08	1.7	58.2	9	4.8
GA5	29/1/99	7.1	30	28	196	69	2.83	25.7	NA	<0.1	1.59	<1.0	<0.1	NA
GA5	18/7/00	7	48	90	224	95	NA	33	NA	NA	2	23	NA	0.2
GA5	28/2/01	7.3	61	22	194	75	NA	25	NA	NA	2	25	NA	0.2
GA17	4/4/98	7.66	65	39	200	56.6	ND	18.2	ND	0.05	4	32.5	10	11.8
GA17	29/1/99	7	165	31	178	93.7	2.13	31.2	NA	<0.1	1.88	25	<0.1	NA
GA17	18/7/00	7	22	34	230	72	NA	25	NA	NA	3	21	NA	0.6
GA17	28/2/01	NA	NA	NA	NA	NA	NA	NA	NA	NA	NA	NA	NA	NA
GA15	29/1/99	7	41	220	248	119	4.28	52.1	NA	<0.1	1.12	33	<0.1	NA
GA15	18/7/00	7.2	56	310	172	117	NA	46	NA	NA	3	67	NA	ND
GA15	28/2/01	NA	NA	NA	NA	NA	NA	NA	NA	NA	NA	NA	NA	NA
GA21	26/8/94	NA	NA	NA	NA	NA	NA	NA	NA	NA	NA	NA	NA	NA
GA21	18/4/96	7.8	163	59	370	131.5	ND	65.5	0.2	0.03	3	53	<0.1	14.5
GA21	17/7/96	7.61	99	70	440	105	0.16	55.8	<0.05	0.05	3	40	NA	0.5
GA21(A)	24/10/96	7.56	139	71	370	107.9	1.4	56.1	0.05	0.03	5	27	NA	5.1
GA21(B)	24/10/96													
GA21	21/1/97	7.54	57	65	360	113	0.31	57	<0.05	0.14	4	32	NA	8.8
GA21	12/8/97	7.4	58	77	350	96.7	<0.05	49.6	<0.05	0.02	<3	27	<0.1	6.1
GA21	4/4/98	7.65	28	57	320	77.9	ND	38.1	0	0.03	2.3	9.5	18	8.1
GA21	29/1/99	7	21	46	272	187	4.84	68	NA	<0.1	1.71	18.3	<0.1	NA
GA21	18/7/00	NS	NS	NS	NS	NS	NS	NS	NS	NS	NS	NS	NS	NS
GA21	28/2/01	NA	NA	NA	NA	NA	NA	NA	NA	NA	NA	NA	NA	NA
GA19	26/8/94	7.4	260	47	392	145	0.03	72	0.07	NA	2.6	32	NA	13
GA19	18/4/96	NA	NA	NA	NA	NA	NA	NA	NA	NA	NA	NA	NA	NA
GA19	17/7/96	NA	NA	NA	NA	NA	NA	NA	NA	NA	NA	NA	NA	NA
GA19	21/1/97	7.6	53	43	260	75	0.13	29.5	<0.05	0.35	6	32	NA	6.6
GA19	18/7/00	NS	NS	NS	NS	NS	NS	NS	NS	NS	NS	NS	NS	NS
GA19	28/2/01	7.5	40	32	168	50	NA	18	NA	NA	3	33	NA	ND
GA19(A)	24/10/96	7.44	81	71	350	97.5	1.06	46	<0.05	0.03	5	25	NA	4.5
GA19(A)	12/8/97	7.42	44	61	350	83.7	<0.05	38.1	<0.05	0.01	<3	50	<0.1	3.2
GA19(A)	4/4/98	7.59	33	40	330	77	ND	33.7	ND	0.04	3.1	15.1	14	5.2
GA19(A)	29/1/99	7	39.5	75	296	86.6	1.97	35	NA	<0.1	1.87	20.4	<0.1	NA
GA19(A)	18/7/00	NS	NS	NS	NS	NS	NS	NS	NS	NS	NS	NS	NS	NS
GA19(A)	28/2/01	7.3	29	28	192	62	NA	25	NA	NA	4	23	NA	0
GA19(B)	24/10/96	NA	NA	NA	NA	NA	NA	NA	NA	NA	NA	NA	NA	NA
GA19(B)	12/8/97	7.56	43	112	290	68.9	<0.05	31.7	0.12	0.02	<3	38	<0.1	2.7
GA19(B)	4/4/98	7.67	51	124	270	68	ND	32.4	0.08	0.11	3	49.5	39	4.1
GA19(B)	29/1/99	7	97	210	310	90.7	0.56	48.9	NA	<0.1	2.57	58.8	<0.1	NA
GA19(B)	18/7/00	NS	NS	NS	NS	NS	NS	NS	NS	NS	NS	NS	NS	NS
GA19(B)	28/2/01	NS	NS	NS	NS	NS	NS	NS	NS	NS	NS	NS	NS	NS

Note NS = location Not Sampled; ND = parameter Not Detected; NA = parameter Not Analysed

INORGANIC PARAMETERS

UPGRADIENT - continued

BH	Date	pH	Chloride mg/L	Sulphate mg/L	Alkalinity mg/L	Ca mg/L	Fe mg/L	Mg mg/L	Mn mg/L	Sulphide mg/L	K mg/L	Na mg/L	S (free) mg/L	Nitrate mg/L
GA7	19/3/94	7.9	51	34	420	81	NA	41	NA	NA	4.6	35	NA	NA
GA7	26/8/94	7.6	46	32	445	170	0.03	52	0.28	NA	6.3	22	NA	13
GA7	18/4/96	7.9	79	54	350	105	0.48	51.7	<0.05	0.03	8	41	<0.1	0.4
GA7	17/7/96	7.95	58	54	380	77.9	0.16	35.7	<0.05	0.02	9	52	NA	2.7
GA7(dup)	24/10/96	7.42	66	68	360	100	1.12	46.7	<0.05	0.03	10	23	NA	4.7
GA7	24/10/96	NA	NA	NA	NA	NA	NA	NA	NA	NA	NA	NA	NA	NA
GA7	21/1/97	7.31	44	49	370	109	0.08	44.4	<0.05	0.07	8	28	NA	10
GA7(dup)	22/2/97	NA	NA	NA	NA	NA	NA	NA	NA	NA	NA	NA	NA	NA
GA7	12/8/97	7.58	30	52	260	70	0.06	25	<0.05	0.02	10	41	<0.1	3
GA7	4/4/98	7.47	43	68	310	84.7	ND	35.3	ND	0.07	3.2	18.5	15	11.6
GA7	29/1/99	7.6	38	75	208	63.3	2.87	25.2	NA	0.48	4.32	21.1	<0.1	NA
GA7	18/7/00	6.5	30	49	212	65	NA	26	NA	NA	4	20	NA	0.6
GA7	28/2/01	7.0	55	39	148	55	NA	19	NA	NA	5	42	NA	1.1
BH19	19/3/94	NA	140	50	350	115	NA	50	NA	NA	3.9	22	NA	NA
BH19	26/8/94	7.3	87	44	445	150	0.09	64	0.25	NA	4.5	23	NA	13
BH19	18/4/96	NA	NA	NA	NA	NA	NA	NA	NA	NA	NA	NA	NA	NA
BH19	17/7/96	NA	NA	NA	NA	NA	NA	NA	NA	NA	NA	NA	NA	NA
BH19	24/10/96	7.64	46	66	210	58.1	1.57	25.6	<0.05	0.03	4	17	NA	1.9
BH19	21/1/97	7.49	71	76	240	75.9	0.08	35.2	0.17	0.03	4	38	NA	6.5
BH19	12/8/97	7.42	39	55	330	79.1	<0.05	35.8	<0.05	0.01	19	35	<0.1	3.2
BH19	4/4/98	7.54	40	47	340	69.7	ND	30	ND	0.03	4.1	33.5	11	4.6
BH19	29/1/99	7.4	21	43	268	60.9	0.4	25.3	NA	<0.1	3.43	18	<0.1	NA
BH19 (dup)	29/1/99	7.4	20.5	43	264	65.5	0.38	26.4	NA	<0.1	3.78	22	<0.1	NA
BH19	18/7/00	6.7	32	30	212	58	NA	26	NA	NA	3	19	NA	0.4
BH19	28/2/01	NA	NA	NA	NA	NA	NA	NA	NA	NA	NA	NA	NA	NA
MWU	18/4/96	8.1	55	60	230	87.1	0.43	21.6	0.06	0.03	<3	49	<0.1	5
MWU	17/7/96	7.55	37	50	220	21.8	1.11	5.79	0.06	0.05	3	74	NA	5.3
MWU	24/10/96	7.77	56	62	270	88.6	1.2	25.5	0.2	0.03	8	29	NA	1.1
MWU	21/1/97	7.71	59	44	340	107	0.08	28.7	0.2	0.05	5	36	NA	5.3
MWU	12/8/97	7.27	24	35	190	47.7	<0.05	14.1	0.17	0.02	<3	18	<0.1	4.6
MWU	4/4/98	7.55	41	34	300	71.4	ND	23.8	0.1	0.02	3.8	31.5	12	6.8
MWU	29/1/99	7.6	30.5	37	138	37.9	3.86	16.5	NA	<0.1	2.29	12.2	<0.1	NA
MWU	18/7/00	6.3	15	17	96	27	NA	8	NA	NA	2	14	NA	ND
MWU	28/2/01	NA	NA	NA	NA	NA	NA	NA	NA	NA	NA	NA	NA	NA

Note NS = location Not Sampled; ND = parameter Not Detected; NA = parameter Not Analysed

INORGANIC PARAMETERS

WITHIN REACTOR

BH	Date	pH	Chloride mg/L	Sulphate mg/L	Alkalinity mg/L	Ca mg/L	Fe mg/L	Mg mg/L	Mn mg/L	Sulphide mg/L	K mg/L	Na mg/L	S (free) mg/L	Nitrate mg/L
R5	18/4/96	8.13	61	57	220	73.6	0.73	20	0.13	0.03	3	61	<0.1	5.9
R5	17/7/96	8.19	51	13	230	42.1	0.13	19.5	<0.05	0.02	5	56	NA	<0.5
R5	24/10/96	8.11	63	30	180	36.8	4.4	21.3	0.18	0.03	5	34	NA	<0.5
R5	21/1/97	NA	NA	NA	NA	NA	NA	NA	NA	0.02	NA	NA	NA	NA
R5	12/8/97	8.34	89	16	60	27	0.08	0.16	<0.05	0.03	<3	39	<0.1	4.1
R5	4/4/98	NA	NA	NA	NA	NA	NA	NA	NA	NA	NA	NA	NA	NA
R5	29/1/99	8	38.5	40	146	37.6	6.71	15.2	NA	0.31	<1.0	8.85	<0.1	NA
R5	18/7/00	7.2	28	11	76	12	NA	7	NA	NA	3	29	NA	0.1
R5	28/2/01	NA	NA	NA	NA	NA	NA	NA	NA	NA	NA	NA	NA	NA
R4	18/4/96	7.99	125	26	30	27.7	0.44	12.8	0.13	0.03	3	39	<0.1	8
R4	17/7/96	8.71	104	11	60	23.4	3.13	10.3	0.06	0.05	5	52	NA	6.6
R4	24/10/96	9.29	98	17	40	14.6	6.13	6.03	0.1	0.04	4	35	NA	<0.5
R4 (dup)	24/10/96	NA	NA	NA	NA	NA	NA	NA	NA	NA	NA	NA	NA	NA
R4	21/1/97	9.19	85	<3	70	12.2	0.05	9.58	<0.05	0.03	5	36	NA	9.8
R4	12/8/97	7.92	87	11	110	23.4	<0.05	6.84	<0.05	0.03	12	50	<0.1	4.3
R4	4/4/98	NA	NA	NA	NA	NA	NA	NA	NA	NA	NA	NA	NA	NA
R4	29/1/99	8.9	47.5	<5.0	44	6.4	2.1	11.5	NA	0.15	1.97	20.6	<0.1	NA
R4	18/7/00	6.9	34	ND	36	5	NA	3	NA	NA	4	27	NA	0.1
R4	28/2/01	NA	NA	NA	NA	NA	NA	NA	NA	NA	NA	NA	NA	NA
R3	18/4/96	9.16	77	6	40	13.3	0.53	0.43	0.05	0.05	4	50	<0.1	3
R3	17/7/96	9	109	8	50	62.1	0.68	19.7	0.14	0.08	6	56	NA	7.8
R3	24/10/96	9.47	107	17	30	17.9	0.49	2.36	<0.05	0.07	6	39	NA	<0.5
R3	21/1/97	9.38	94	<3	40	15.9	1.08	1.32	<0.05	0.03	5	41	NA	7.3
R3	12/8/97	7.97	85	14	80	21.3	<0.05	1.75	<0.05	0.03	<3	38	<0.1	3.1
R3	4/4/98	NA	NA	NA	NA	NA	NA	NA	NA	NA	NA	NA	NA	NA
R3	29/1/99	8.9	87	<5.0	36	19.6	1.52	8.92	NA	<0.1	1.26	28.5	<0.1	NA
R3	18/7/00	7.2	57	0	36	12	NA	2	NA	NA	4	38	NA	0.1
R3	28/2/01	NA	NA	NA	NA	NA	NA	NA	NA	NA	NA	NA	NA	NA
R2	18/4/96	9.2	55	6	50	4.95	0.58	0.5	<0.05	0.05	4	50	<0.1	2.4
R2	17/7/96	9.58	116	3	50	29.3	1.1	0.84	<0.05	0.05	5	52	NA	7.8
R2	24/10/96	9.21	102	32	50	25.4	0.46	5.56	<0.05	0.02	6	44	NA	<0.5
R2	21/1/97	9	88	<3	60	12.4	0.18	2.14	<0.05	0.03	5	41	NA	5.8
R2	12/8/97	7.22	66	17	100	11	<0.05	11.66	<0.05	0.02	<3	35	<0.1	3.2
R2	4/4/98	NA	NA	NA	NA	NA	NA	NA	NA	NA	NA	NA	NA	NA
R2	29/1/99	9	68	<5.0	42	18.2	2.35	5.11	NA	<0.1	3.51	32.8	<0.1	NA
R2	18/7/00	7.7	76	ND	36	16	NA	3	NA	NA	5	41	NA	0.2
R2	28/2/01	NA	NA	NA	NA	NA	NA	NA	NA	NA	NA	NA	NA	NA
R1	18/4/96	9.3	43	24	50	5.75	1.08	0.52	<0.05	0.04	5	59	<0.1	2.6
R1	17/7/96	9.67	46	22	110	6.93	0.09	0.26	<0.05	0.08	6	55	NA	<0.5
R1	24/10/96	9.91	65	22	50	8.94	1.31	1.26	<0.05	0.02	10	46	NA	<0.5
R1	21/1/97	9.83	83	8	50	17.3	0.8	0.14	<0.05	0.04	6	43	NA	8.3
R1(dup)	22/1/97	NS	NA	NA	NA	NA	NA	NA	NA	NA	NA	NA	NA	NA
R1	12/8/97	7.92	32	21	190	23.6	0.07	19.58	<0.05	0.03	19	43	<0.1	5.5
R1	4/4/98	NA	NA	NA	NA	NA	NA	NA	NA	NA	NA	NA	NA	NA
R1	29/1/99	9.8	46.5	<5.0	46	16.1	2.51	0.348	NA	<0.1	4.16	35.6	<0.1	NA
R1	18/7/00	7.7	65	ND	34	25	NA	ND	NA	NA	4	37	NA	0.1
R1	28/2/01	NA	NA	NA	NA	NA	NA	NA	NA	NA	NA	NA	NA	NA

Note NS = location Not Sampled; ND = parameter Not Detected; NA = parameter Not Analysed

INORGANIC PARAMETERS

DOWNGRADIENT

BH	Date	pH	Chloride mg/L	Sulphate mg/L	Alkalinity mg/L	Ca mg/L	Fe mg/L	Mg mg/L	Mn mg/L	Sulphide mg/L	K mg/L	Na mg/L	S (free) mg/L	Nitrate mg/L
MWD	18/4/96	11.18	99	92	250	28.3	1.91	1.11	0.06	0.02	3	202	<0.1	8
MWD	17/7/96	10.59	85	123	180	5.97	0.3	0.26	<0.05	0.03	8	165	NA	10.4
MWD	24/10/96	8.89	70	48	130	30.5	2.17	2.63	0.07	0.1	<3	55	NA	1.7
MWD	21/1/97	8.62	291	18	110	93.2	0.38	6.31	0.07	0.03	11	98	NA	6.7
MWD	12/8/97	8.45	80	45	140	40.4	<0.05	4.11	0.06	0.02	<3	45	<0.1	4.1
MWD	4/4/98	NA	NA	NA	NA	NA	NA	NA	NA	NA	NA	NA	NA	NA
MWD	29/1/99	7.9	288	17	86.1	73.3	3.24	19.8	NA	0.64	2.51	73.5	<0.1	NA
MWD	18/7/00	7.3	51	38	100	33	NA	7	NA	NA	1	53	NA	0.2
MWD	28/2/01	NS	NS	NS	NS	NS	NS	NS	NS	NS	NS	NS	NS	NS
GA12	19/3/94	NA	56	55	430	105	NA	46	NA	NA	1.8	16	NA	NA
GA12	26/8/94	8.7	41	52	592	110	<.01	39	0.02	NA	4.3	210	NA	12
GA12	18/4/96	NA	NA	NA	NA	NA	NA	NA	NA	NA	NA	NA	NA	NA
GA12	17/7/96	7.7	34	63	430	164	0.61	43.5	<0.05	0.03	6	56	NA	5.5
GA12	24/10/96	Broken	Sample											
GA12	21/1/97	7.61	27	58	340	97.3	0.14	44.7	<0.05	0.21	4	24	NA	8.6
GA12	12/8/97	7.44	23	78	400	91.2	<0.05	43.8	<0.05	0.02	<3	29	<0.1	1.1
GA12	4/4/98	7.68	36	70	310	84.5	ND	34.9	ND	0.13	2	11.5	18	2.8
GA12	29/1/99	7.8	25.5	40	220	162	1.64	27.4	NA	<0.1	3.29	<1.0	<0.1	NA
GA12	18/7/00	7.1	28	44	274	80	NA	38	NA	NA	2	37	NA	0.1
GA12	28/2/01	NA	NA	NA	NA	NA	NA	NA	NA	NA	NA	NA	NA	NA
GA13	19/3/94	NA	62	56	430	90	NA	51	NA	NA	3.5	30	NA	NA
GA13	26/8/94	7.7	97	325	357	170	0.77	78	0.59	NA	3.8	49	NA	10
GA13	18/4/96	7.79	163	87	360	147.4	0.13	74.1	0.07	0.03	8	36	<0.1	4.6
GA13	17/7/96	7.97	158	50	370	145	0.35	63.6	<0.05	0.03	8	37	NA	<0.5
GA13(A)	24/10/96	7.5	187	201	340	137.7	0.89	71.3	<0.05	0.03	8	34	NA	1.1
GA13(A)	24/10/96	NA	NA	NA	NA	NA	NA	NA	NA	NA	NA	NA	NA	NA
GA13	21/1/97	7.37	149	112	360	180	0.11	78.7	0.06	0.13	10	36	NA	7.3
GA13	12/8/97	7.41	126	292	370	162.4	<0.05	70.55	0.05	0.02	14	47	<0.1	5.5
GA13	4/4/98	7.75	112	120	290	117	0	47.4	ND	0.22	2.9	20	35	6.3
GA13	29/1/99	7.2	75.5	41	323	69.8	1.1	34.6	NA	<0.1	5.9	21.6	<0.1	NA
GA13	18/7/00	NS	NS	NS	NS	NS	NS	NS	NS	NS	NS	NS	NS	NS
GA13	28/2/01	7.3	95	28	266	95	NA	45	NA	NA	5	25	NA	0.5
GA14	19/3/94	NA	NA	NA	NA	NA	NA	NA	NA	NA	NA	NA	NA	NA
GA14	26/8/94	NA	NA	NA	NA	NA	NA	NA	NA	NA	NA	NA	NA	NA
GA14	18/4/96	10.73	495	81	60	75.7	5.95	0.49	0.08	0.02	9	421	<0.1	16.6
GA14	17/7/96	9.4	449	91	90	50.4	<0.05	1.56	<0.05	0.05	6	315	NA	10.1
GA14	24/10/96	10.23	417	120	40	43.4	1.23	1.91	<0.05	0.02	5	204	NA	11.6
GA14	21/1/97	10.31	410	89	70	80.4	0.1	2.4	<0.05	0.09	9	229	NA	16.7
GA14(dup)	21/1/97	NA	NA	NA	NA	NA	NA	NA	NA	NA	NA	NA	NA	NA
GA14	4/4/98	9.88	347	67	120	57.8	ND	4.04	ND	0.02	3.1	201	25	17
GA14	29/1/99	10.4	410	90	<2.0	155	0.75	12.7	NA	<0.1	9.18	207	<0.1	NA
GA14	18/7/00	NA	NA	NA	NA	NA	NA	NA	NA	NA	NA	NA	NA	NA
GA14	28/2/01	NA	NA	NA	NA	NA	NA	NA	NA	NA	NA	NA	NA	NA

Note NS = location Not Sampled; ND = parameter Not Detected; NA = parameter Not Analysed

DOWNGRADIENT - continued

INORGANIC PARAMETERS

BH	Date	pH	Chloride mg/L	Sulphate mg/L	Alkalinity mg/L	Ca mg/L	Fe mg/L	Mg mg/L	Mn mg/L	Sulphide mg/L	K mg/L	Na mg/L	S (free) mg/L	Nitrate mg/L
GA6	19/3/94	7.5	310	150	75	150	NA	82	NA	NA	3	56	NA	NA
GA6	26/8/94	7.6	245	450	238	190	0.54	92	1.06	NA	4.2	99	NA	10
GA6	18/4/96	7.98	265	146	210	137.7	0.14	79.5	<0.05	0.03	4	70	<0.1	7
GA6	17/7/96	8.06	288	141	230	124	0.2	72.2	<0.05	0.07	5	80	NA	<0.5
GA6	24/10/96	7.34	293	225	240	154.2	0.13	84.9	0.41	0.03	7	74	NA	<0.5
GA6	21/1/97	7.4	236	148	300	138	<0.05	70.4	0.14	0.08	6	80	NA	7.3
GA6	12/8/97	7.39	210	217	330	125.2	0.06	67.1	0.78	0.01	17	41	<0.1	3.4
GA6	4/4/98	6.95	166	257	340	134	0.09	58.7	0.8	0.02	4.2	85.7	72	4.3
GA6	29/1/99	NS	NS	NS	NS	NS	NS	NS	NS	NS	NS	NS	NS	NS
GA6	18/7/00	NS	NS	NS	NS	NS	NS	NS	NS	NS	NS	NS	NS	NS
GA6	28/2/01	NA	NA	NA	NA	NA	NA	NA	NA	NA	NA	NA	NA	NA
GA10	26/8/94	7.8	20	170	317	170	0.81	66	0.7	NA	2.6	30	NA	10
GA10	18/4/96	NA	NA	NA	NA	NA	NA	NA	NA	NA	NA	NA	NA	NA
GA10	17/7/96	8.22	40	140	450	93.5	<0.05	55.5	<0.05	0.02	6	51	NA	<0.5
GA10	24/10/96	7.67	44	154	350	96.3	1.1	55.2	0.08	0.03	4	30	NA	<0.5
GA10	21/1/97	NA	NA	NA	NA	NA	NA	NA	NA	NA	NA	NA	NA	NA
GA10	12/8/97	NA	NA	NA	NA	NA	NA	NA	NA	NA	NA	NA	NA	NA
GA10	4/4/98	7.5	18	81	300	74.2	ND	26.1	ND	0.11	2.9	28	23	10.8
GA10	29/1/99	NS	NS	NS	NS	NS	NS	NS	NS	NS	NS	NS	NS	NS
GA10	18/7/00	NS	NS	NS	NS	NS	NS	NS	NS	NS	NS	NS	NS	NS
GA10	28/2/01	NA	NA	NA	NA	NA	NA	NA	NA	NA	NA	NA	NA	NA

Note NS = location Not Sampled; ND = parameter Not Detected; NA = parameter Not Analysed